RedShift

College Edition
Astronomy Workbook

Second Edition

Bill O. Walker

Late of Tyler Junior College

THOMSON

BROOKS/COLE

Australia • Canada • Mexico • Singapore • Spain • United Kingdom • United States

Cover design and digital illustration, Bob Western.

Photo of head with glasses, courtesy of Corbis Images.

Trifid Nebula photo, courtesy of 2MASS (Atlas Image mosaic obtained as part of the Two Micron All Sky Survey, a joint project of the University of Massachusetts and the Infrared Processing and Analysis Center/California Institute of Technology, funded by the National Aeronautics and Space Administration and the National Science Foundation.)

Screen shots on the cover image are courtesy of Maris Multimedia.

Printed in Canada
2 3 4 5 6 7 08 07 06

Printer: Webcom Limited

ISBN: 0-534-49031-X

Thomson Brooks/Cole
10 Davis Drive
Belmont, CA 94002-3098
USA

Asia
Thomson Learning
5 Shenton Way #01-01
UIC Building
Singapore 068808

Australia/New Zealand
Thomson Learning
102 Dodds Street
Southbank, Victoria 3006
Australia

Canada
Nelson
1120 Birchmount Road
Toronto, Ontario M1K 5G4
Canada

Europe/Middle East/South Africa
Thomson Learning
High Holborn House
50/51 Bedford Row
London WC1R 4LR
United Kingdom

Latin America
Thomson Learning
Seneca, 53
Colonia Polanco
11560 Mexico D.F.
Mexico

Spain/Portugal
Paraninfo
Calle/Magallanes, 25
28015 Madrid, Spain

I would like to dedicate this book to my wife and helper, Carolyn, whose advice, assistance, and understanding have made this book possible.

Contents

Groups and Galaxies

Appendices: Observing with *RedShift*

THE LAUNCHING PAD

Do not pass GO—Stop here!
Before proceeding, you *must* read this page!

Welcome to the incredible world of *RedShift*: a powerful planetarium, a trusty field companion, and an invaluable educational assistant. In the following pages, you will learn how to use this powerful software package. As you go along, *RedShift's* capabilities will help you understand more about our position in the universe and take you to places that hadn't been dreamed of a hundred years ago.

You Are the Astronomer!

The exercises in this workbook have been designed to lead *you* step by step through *real* astronomical observations using *RedShift's* simulations. These are the same observations the great astronomers have made through the centuries to carve out our present understanding of the universe. The questions have been designed to help you interpret your observations through logical thinking and discover important concepts for yourself. *Each exercise gives you the opportunity to be an astronomer.* You will have to use your powers of observation and then of deduction to assemble, little by little, a comprehensive model of our universe.

Each exercise in this workbook is designed to take you to a specific conceptual destination. Understanding the layout of the exercises will help you reach that destination with a minimum of wasted time and effort. Please take a moment now to familiarize yourself with the exercise format.

I. **Introduction and Purpose:** Each exercise begins with a short introduction to the topic, along with *a specific statement of the purpose of the exercise*. Read this purpose carefully so that you will know just what it is you are supposed to accomplish.

II. **Procedure:** *Always start RedShift from scratch for every new exercise.* If you want to do two exercises in a row, you must close the entire *RedShift* program and then restart it before doing the second exercise. Each step-by-step procedure is accompanied either by a description of what the simulation should look like or by actual photos of a typical computer screen. *If you do not see what you expect*, check to see that you have followed the procedure exactly as written.

1

III. **Data Table:** When appropriate, a table will be included for recording your data so that *you should not have to wonder about what you are supposed to measure*.

IV. **Questions:** The interpretation of your observations and data will be accomplished through a set of *questions that will lead you through the thought processes necessary to understand the concepts* introduced in the Purpose section of the exercise.

V. **Discussion and Conclusion:** Each exercise will conclude with a brief discussion and summary of your observations, and you will be asked to write a short conclusion of your own. Remember, you are the astronomer, and this is *your* report of your observations and findings. Be thorough and specific about what you observed, showing that you accomplished the purpose set out at the beginning.

Needed Equipment

To use *RedShift College Edition,* you must have one of the following systems *at a minimum*:

- An IBM-compatible computer with a Pentium processor (90 MHz or faster), 32 megabytes of RAM, 4× CD-ROM drive, Windows 95 or later, 800 × 600 video card with High Color (or True Color), and a Windows-compatible sound card.

- A PowerMac 100 (or above) with System 7.61 or later, 4× CD-ROM drive, 32 megabytes of RAM (64 MB recommended), 800 × 600 Resolution in High Color (or above).

The universe is waiting! Have fun!

I
Lift-Off!

Contents

And Now: Heeeere's *RedShift!*

Introduction and Purpose

The purpose of today's exercise is to introduce you to the *RedShift* software package. You will learn how to start *RedShift*, how to select controls and panels from the menus, and how to use many of *RedShift's* different kinds of buttons. We have tried to make the instructions as clear as possible, but there may still be things that don't seem to work. Don't be afraid to ask your instructor for help if you get stuck.

In this exercise, you will no doubt encounter many unfamiliar things. *RedShift* is a very powerful and complex collection of tools for learning astronomy; it cannot be mastered in one or two hours. So, in today's exercise, don't worry about specific terms or try to remember where every control is or what every button stands for. The main goal is to get you comfortable with using *RedShift*. We have plenty of exercises designed to work out the specifics later. So, take a deep breath and fire up the computer—your own private *RedShift* Universe is waiting!

Important Note: This *RedShift* software package is supplied on a CD-ROM disc. You *must* have a computer with a CD-ROM drive in order to use this software. You will want to install the main *RedShift* program on your hard drive first before using the software.

Installing *RedShift College Edition*

Windows:

1. Insert the *RedShift* CD-ROM in the CD-ROM drive of the computer and close the CD-ROM drive door. After a few seconds, the CD will spin and a small dialog box will open stating that this application has not been installed on your computer. Click **INSTALL** to begin the installation.

2. Indicate where you want the software installed. You may install the software on any hard drive in your system, but the default location is probably the best. Click **NEXT** to create a new directory on your C: drive or **BROWSE** to select another location or drive.

3. Install ancillary software needed to run the *RedShift* software. Current versions of both Microsoft's Internet Explorer and Apple's QuickTime

software are necessary to use *RedShift College Edition*. If your computer is new, you might have a more recent version of Internet Explorer already installed, so watch for such warnings. If this is the case, skip the installation of Internet Explorer. When you have finished selecting (or deselecting) the appropriate software, click on the **NEXT** button.

4. Now the *RedShift College Edition* files will be installed. When all the necessary *RedShift* files have been copied to your computer's hard drive, a small window will appear on your screen stating that *RedShift 3* has been successfully installed. Click **OK**.

5. Finally, you will be asked if you wish to install QuickTime 3.0. If you have a higher version of QuickTime already installed on your computer, click **NO**; otherwise, click **YES**. Respond to the prompts that follow as is appropriate for your computer system. When the license agreement is displayed, read the terms of the license and then click **AGREE** to signify your compliance. A small window will open to lead you through the QuickTime installation.

6. Before finishing the installation, select the **PLAY MOVIE** option to confirm that the QuickTime software is working properly on your computer. This command will open a small Movie Player window titled "Sample." Click on the small VCR-style play button at the bottom of the small window to play the QuickTime movie sample. If the movie plays accompanied by some sound effects, then your installation has been successful. Now close the Movie Player window. When all the QuickTime files have been installed, close the QuickTime file folder. Now you are ready to skip to Section 1.1 to begin your *RedShift* session.

Macintosh:

1. Quit all open applications.

2. Load the *RedShift 3* CD-ROM into the CD-ROM drive of the computer and close the drive door. After a few seconds the drive will spin, the *RedShift 3* icon will appear on the Desktop, and a folder window will open. Double-click on the icon marked **RS3 Installer** to begin installation. When the dialog box opens, click **Continue**.

3. Indicate where you want the software installed. You may install the software on any hard drive in your system, but the default location is probably the best. Click **Select Folder...** if you want to select another

location or drive. Once you have selected a location (or decided to use the default location), click **Install** to begin the installation.

4. Now the *RedShift* software files will be installed. When all the necessary *RedShift* files have been copied to your computer's hard drive, another dialog box will open. Click **Quit**.

5. Your Macintosh computer will already have a version of QuickTime installed on it. The *RedShift* CD-ROM has version 3.0 of this software on it. If your QuickTime software is older, you may wish to update it. If you choose to update your files, *you will have to restart your computer when you finish installing these files.* If you wish to update your QuickTime software to version 3.0, continue with step 6; otherwise, skip to Section 1.1.

6. To install version 3.0 of Apple's QuickTime software, look through the folder and click on the icon marked **Install QuickTime…** A dialog box will open to ask if you wish to delete the old QuickTime files before installing the new ones. If you do not wish to delete the old QuickTime files, click **Cancel** to leave your QuickTime installation as it is and go on to Section 1.1. Otherwise, click **Continue** to install the version 3.0 files.

A Welcome window will open, displaying some information about QuickTime. After you have finished reading the information, click **Continue**. A window will open displaying Apple's Software License Agreement. After reading the License Agreement, click **Agree**.

The next step asks on which drive you want the software installed. Select your hard drive and click **Install** to begin the installation. When all necessary files have been copied to your computer's hard drive, a dialog box will appear. Click **Restart** to reset your computer so it will recognize the new QuickTime files.

1

Procedure

Section 1.1

Windows:

Once you have completed the installation of the *RedShift* software, open the CD door and then close it again. After a few seconds, the CD-ROM drive should spin, and a small window should appear on your screen, titled *RedShift College Edition*. Click on the **START** button.

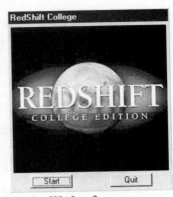

Macintosh:

Once you have completed the installation of the *RedShift* software, open the CD door and then close it again. After a few seconds, the CD-ROM drive should spin, and a small folder with several icons should open on the desktop. In this folder, locate the icon titled *RedShift*. (Scroll around in the folder if necessary.) Double-click on this icon to start *RedShift 3*.

Section 1.2

A small screen should now appear stating that *RedShift 3* is loading. This screen will be followed by the Brooks/Cole logo and the *RedShift 3* title screen, all accompanied by a musical fanfare. Finally, the *RedShift 3* window and then the *RedShift College Edition* window will open on your screen.

Before you go any farther, let's take a short overview of this powerful software suite. The *RedShift College Edition* package you have just installed consists of two separate programs—a special version of *RedShift 3* and *RedShift College Edition*. The specially modified version of *RedShift 3* contains the main features of the standard *RedShift 3* software package plus a special set of tools for making measurements on the *RedShift* sky. *RedShift College Edition* contains an interactive, narrated tutorial called "The Science of Astronomy" and functions as a hub from which you can access the World Wide Web and most of *RedShift 3's* amazing simulations, photos, and prediction tools.

The software automatically opens the *RedShift College Edition* window on top. However, the exercises in this book will work directly with *RedShift 3* most of the time, so you will have to switch to the *RedShift 3* window each time you start a new exercise. You can switch back and forth between the two windows at any time by clicking on the **Window** menu at the very top of the screen and selecting *RedShift 3/RedShift College Edition*.

Section 1.3

If this is your first time to start *RedShift College Edition*, you will need to set some initial parameters so the software will make proper calculations for your viewing location. The Preferences window should open automatically the first time *RedShift College Edition* is started. (If it does not, select *RedShift 3/RedShift College Edition* from the **Window** Menu at the top of the screen. This will hide the *RedShift College Edition* window and reveal the *RedShift 3* window. Now select **Preferences** from the **File** Menu to open the Preferences window.)

There are several parts to this window that need to be set. First, notice the **date** format and **time** format. For the exercises in this book, *these two parameters should be set as shown here.* Date: **MM/DD/YYYY** and Time: **am/pm**. If either parameter is different than shown, click on the parameter and select the appropriate setting from the menu that will drop down.

Once these two parameters are set properly, you will need to set the location to your home location. (The location shown above is for Tyler, Texas.) If you know the longitude and latitude of your viewing location, you may enter the numbers directly into these boxes; otherwise, you will need to click on the command **Set on Map**.

Clicking on this button opens the Choose Location window as shown here. In this window, you will first need to place the cursor over the small globe. Hold the left mouse button down and rotate the globe until North America is centered. Now move the mouse until the cursor is over the yellow circle.

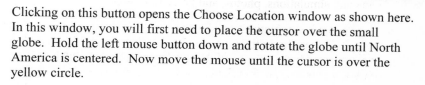

9

1

When the cursor is centered on the yellow circle, click the left mouse button, and the circle will turn red. Hold the mouse button down, drag the circle until it is approximately over your home location, and then release the button. The yellow crosshair on the upper map shows a more exact position.

Now click the Magnify button just below the map window. This will magnify the region around the crosshair and put city names on the map. Place the cursor approximately where you think your viewing location is and double-click the left mouse button to move the crosshair to this spot. When you have the yellow crosshair set to the approximate location of your viewing site, click **OK** at the bottom of the window to close the Choose Location window. Finally, click **OK** at the bottom of the Preferences window to finish up.

Section 1.4

Now that you have set your location, let's take a look around the basic *RedShift 3* screen. If you are still looking at the *RedShift College Edition* title screen, click on the **Window** menu and select ***RedShift 3/RedShift College Edition***. At the top of the screen is a menu bar with nine menu titles:

File Edit Display Control Tools Information Events Window Help

These titles represent drop-down menus that contain most of *RedShift 3's* options and controls. They can be opened in one of two ways. You can open a menu by clicking on it with the mouse or from the keyboard by holding down the <Alt> key and pressing the letter that is underlined in each menu title (i.e., <Alt>"F" opens the File menu). To select a command from a menu, click the left mouse button while the cursor is positioned over it. To close a menu without selecting anything from it, click on the menu title again. Find the answers to the following questions by opening and exploring *RedShift's* menus.

Q1 Open each menu and find the one containing an entry titled **Status Bar**. Which menu was this entry listed under?

The Status Bar option should be selected with a check mark in front of it. Click on **Status Bar** to turn this option **Off**, and notice what part of the screen disappears. (*Hint:* Look down.) Now open the menu again, click on **Status Bar** once more to turn the bar back **On**, and notice the information displayed here.

Q2   In the same menu is an entry titled **Panels**. This entry is also checked. Click on the **Panels** entry and see what part of the screen disappears. Now use the menus to turn the Panels option back **On**.

There are two different sets of panels—the Activities (or Control) panels and the Settings panels. There are numerous buttons and data boxes on these panels. Click on any of the blank areas of the panels, hold the left mouse button down and drag the mouse. What happens to the panel when you perform this action?

By dragging the panels on the screen, you can move them so that they don't obscure areas that you might wish to see.

Q3   There are three icons at the top right part of each of the two panels. By clicking on each of these buttons, you can open and close the panels associated with each icon. What is the name of the panel associated with the small clock icon?

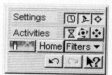

Q4   Which panel is controlled by the middle Settings icon?

Q5   Which panel is controlled by the third Settings icon?

These three Settings panels will be used often in our *RedShift* explorations, so remember how to turn them **On** and **Off**.

Q6   Now turn your attention to the **Activities** (or Control) buttons. What is the name of the panel associated with the first Activities button that looks like an hourglass?

Q7   The Control Time panel is an extremely powerful feature of *RedShift 3*. We will be using this control frequently, so you should get acquainted with its features. First, notice the VCR-style buttons at the top of the panel. These buttons function like those on your VCR: **Reverse Play**, **Reverse Step**, **Stop**, **Forward Step**, and **Play**. Click the **Forward Step** button and watch the screen and the Time settings panel.

1

Each time you click the **Forward Step** button, how many hours pass in the Time settings panel?

Q8   Just below the VCR buttons is a data box titled **Step**. Click on the small arrow to the right of the Step box.

Choose the **Units** option from the first drop-down menu and write down the different options for the units of the Time Step shown in the second drop-down menu.

Q9   Now click on the part of the Step box that contains the number. Change the number either by typing in a new number or by clicking again on the small arrow and choosing a multiply or divide option. What effect will the new setting have on the change in the Time settings panel when the **Forward Step** button is clicked again? Now close the Control Time panel by clicking again on its button in the Activities Panel.

Q10   To which menu would you go to find information about upcoming solar or lunar eclipses or planet conjunctions?

Q11   List the items found in the **Information** menu.

Q12   In the **Window** menu, which items are selected (that is, which have check marks beside them)?

Q13   Below the Settings and Activities Panels is a small area with the panel display icon and buttons titled **Home** and **Filters**. Click on the panel display icon 🔲 to the left of the Home button and then select each option in the icon menu that drops down. What do these selections do?

Q14   Now click on the **Filters** button. What choices are listed under the Filters drop-down menu?

Q15   Select **Constellations** from the Filters menu. In the Filters window, you will see abbreviations of the names of all the official constellations. You can display all or some of them from this window by clicking on the box next to each constellation to turn its check mark on or off.

In the following box, fill in the blank spaces, and then calculate the total number of constellations listed.

Number of constellations in the first column	=	
Number of columns	=	
Multiply the two numbers	×	
Total number of constellations		

Q16 From this window, you can also choose to change the way the selected constellations are displayed. Make sure that the boxes titled **Patterns** and **Names** are checked. Finally, click on the **Constellations** box at the left edge of the window to turn the Constellation feature **On**, and click the **OK** button to see the effect.

Now re-open the Constellation Filters window. This time, add the **Boundaries** option and click **OK**. How are boundaries different from patterns?

Q17 Look through the menus at the top of the screen for the **Find** function. In which menu is the **Find** function?

Q18 Select the **Find** option. In the Find Object window, notice the **Find** selection box at the top. Click on **All**, and then, from the menu that drops down, select **Planets** to change the listing in the Find List from **All** to **Planets**. Now scroll down through the Find List to **Jupiter**. Double-click on **Jupiter** with the left mouse button to select the planet and place several options in the **List Links** Box. Click on **Information on Jupiter**, and then click the **OK** button to open Jupiter's information box.

Most basic information about the planet, such as its position in the sky and its distance from the Earth and Sun, is initially displayed in the box. Additionally, you can find many other statistics and information by clicking on the **More** button. When you have finished looking at Jupiter's statistics, click the **Center** button to rotate the view screen to center on Jupiter, and then click **Close** to close Jupiter's Information box. Is Jupiter above or below your horizon at the present time? (If the sky background is green, it is below the horizon.)

1

Q19 From the Display Menu, click on the item **Natural Sky Color** to remove the check mark and turn off the blue sky if it is daytime. Again from the Display Menu, click on the item **Guides** to remove that check mark and turn off the green Horizon color. Now you should be able to see the stars and constellations all across the screen.

From the Settings panel icons, find the panel titled Aim. The Aim panel determines where the screen is centered and what its magnification will be. Click on the small arrow to the right of the Aim box as shown. From the menu that drops down, select **Deep Sky**. Then, from the Deep Sky menu that appears, select **Orion Nebula** and click with the left mouse button to rotate the screen to center on the Orion Nebula.

Can you see anything at the center of the screen other than stars? You are centered on the Orion Nebula, but you can't see it because it is too small and too dim. Now let's use the Zoom box to change the situation. There are two ways to change the Zoom factor: You can click on the number and type in any value directly, or you can click on the small arrow at the right of the Zoom box and select one of the pre-defined magnifications. Now click on the arrow and select a magnification of **50** to zoom in for a telescopic view of the Orion Nebula.

In a sentence or two, describe the appearance of the Orion Nebula as shown on your screen. This is a fairly realistic view of this amazing cloud of gas and dust as seen through an eight- or ten-inch telescope.

Section 1.5

By now, you have made a lot of changes to the *RedShift* display. Many of the Guided Tours used in *RedShift College Edition* also change the display settings, and once you run one, the sky is hopelessly messed up for "normal" sky-gazing activities. If you should ever change too many parameters or switch to *RedShift* after running a Guided Tour, here is an easy solution. To get the sky back to normal, just choose **Open Default Settings** from the File Menu, and the sky will revert to its original condition.

Section 1.6

When you have returned the sky display to its normal condition by opening the Default settings, click on the Filters button in the panels and select **Stars**. At the top of the Filters window is a horizontal slider labeled **Magnitudes**. The left-hand upper slider determines the brightest objects that will be displayed, and the right-hand upper slider determines the dimmest ones that will be displayed. *(Please note that the slider does not change the brightness of the objects as they appear on the screen. It just decides whether very faint or very bright objects should be shown on the screen or omitted.)*

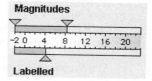

Position the cursor over the right-hand upper slider, and press the left mouse button. The slider should change color. While still holding the left button down, move the mouse to the right until the slider reaches 20 and then release the mouse button. This action of clicking and holding down the left mouse button while moving the mouse is called "dragging" the mouse.

This setting instructs *RedShift* to show all very faint stars (any that are brighter than 20th magnitude). Twentieth-magnitude objects are very faint and cannot be seen with average telescopes! To see what this will look like, click on the **Natural Sky Color** box to remove its check mark and simulate darkness even in daylight hours. In the small window labeled **Show Preview** at the lower left corner of the Filters window, you will see a preview of the results of this action. Now click **OK** to record your selection. What happens to the sky display?

Section 1.7

Now let's turn our attention back to *RedShift College Edition (RSCE)*.

1

From the Window menu, click on **RedShift 3/RedShift College Edition** to re-open the *RSCE* screen. We will be using only a few of *RedShift College Edition's* many valuable features, but you need to be familiar with some of the components of this screen. First, notice the navigation bar that extends all the way across the top of the screen.

We will be using four of these buttons frequently. The button at the left end of the bar, **Contents**, opens and closes the Contents frame at the left side of the RSCE window. Click on it repeatedly to see how it works. When the Contents frame is open, you can select individual chapters in the Science of Astronomy, experiments in the Astronomy Lab, or sections in the Photo Gallery. The large **Prev**, **Go**, and **Next** buttons at the right side of the navigation bar allow you to move around in the Science of Astronomy tutorials. If you do not already know how these buttons work, open the Contents frame and click on **How to Use *RedShift College Edition*** in the Introduction section of the Science of Astronomy.

The Science of Astronomy is not the only resource that can be opened from this screen. Another very valuable tool is the Nightly News (or Sky Diary). Click on the tab labeled **Science of Astronomy**. From the menu that drops down, select the entry titled **Nightly News** to open *RedShift's* powerful Sky Diary window, and answer the following questions.

Q20 The Sky Diary window lists all the important astronomical events for any given month. By default, the window opens to the month corresponding to the date to which *RedShift* is set when the Sky Diary is opened.

Scroll through the Sky Diary window and notice all the different sorts of information included here. When you have looked through the entire month, go back, find the date of the full moon for the current month, and write it here.

Q21 Close the Sky Diary window by clicking on the **Close** button. In the *RSCE* window, again click on the **Science of Astronomy** tab in the navigation bar. This time, select **Photos** to place the Photo Gallery index in the Contents frame. Click on any of the major topics listed to open a secondary list of specific photo collections. Select any subject that interests you to open a list of the photos available on this subject. Scroll down through the list until you find a picture you wish to look at.

Click on the small **Magnify** icon  on the picture to enlarge the picture to full size. When you have finished looking at the picture, click on the **Thumbnail** icon to go back to the window with the picture list in it. Look at as many pictures as you wish. Write a short description of one of the pictures.

Web Intrigues

Q22 *(If you do not have Internet access, you may skip this question.)*

Log on to the Internet and open your web browser. Now turn over to Appendix F and pick a web site that looks interesting. When the browser window opens, type in the URL of the site you chose from Appendix F and check it out. Once you have found an interesting site, write its name and URL here, as well as a short description of what you found there.

Section 1.8

Your *RedShift College Edition* set also contains two exceptional resource collections on the second CD. Push the **Open/Close** button on the CD-ROM drive to open the CD-ROM door. Remove the *RedShift College Edition* Disc 1, insert the **Extras** disc and push the **Open/Close** button once more to load the second CD. When the *RedShift* start screen appears, click **Start**. In a few moments, the *RedShift College Edition Extras* title screen will open in a browser window. *(Note: If you have your screen resolution set at 800 × 600 or lower, the new browser window may open at the bottom of the screen. Click on the top of the browser window and drag the new window up to see it. If you can't see the buttons, scroll down or "pull" the Extras window to a larger size.)*

Once the new window has opened, select **Movie Gallery** from the selection buttons at the bottom of the screen. This will open the Movie Gallery which lists the various movies in the *RedShift* movie archive. Select the movie titled **Moon Roving Vehicle**. Click on the small Film icon to open it.

1

When the movie screen appears, you will see a control bar extending all the way across the bottom from left to right. At the left side of the control bar is a VCR-style button. Click on the various buttons to see what happens. Now watch the movie.

Q23   Close the small movie player window by clicking on the small square ⌧ in the upper right-hand corner of the window to close it. *(Macintosh users will click the small square in the upper left corner of the window to close it.)* Then click the **Back** button at the top of the browser window to go back to the Extras title screen. Now click on the entry titled **The Story of the Universe** to open the list of Story of the Universe segments. From this list, scroll down and select the entry titled **History of the Solar System** to open the movie window, and load a narrated animation describing the formation of the Solar System. As you watch the movie, see if you can discover the length of time astronomers estimate has passed since our Sun began to shine. Write the answer here.

Discussion

In today's exercise, you have gotten just a glimpse of the power and the exciting capabilities of the various components of *RedShift College Edition*. If controlling so many different tools seems overwhelming to you right now, don't worry! We have plenty of time to give each one our undivided attention. By the time you've done all the exercises in this manual, you'll be a *RedShift* pro!

You should be familiar enough now, however, with the various menus and selection methods that you can navigate the procedures in future exercises without too much difficulty. In the exercises that follow, *RedShift's* power will launch you into the unknown to discover the magic and beauty of our incredible Universe!

Conclusion

In this exercise, you have tried out most of the different kinds of interaction needed to use *RedShift*. You have used on/off buttons, check-mark boxes, data entry boxes, sliders, and VCR controls. Beside each of the interactions listed on the next page, give one specific example of a *RedShift* control or function that uses that type of interaction. The first one is done for you as an example.

Type of Control	Where It Is Used in *RedShift*
On/off buttons	**the icon buttons in the panels (buttons toggle on and off)**
Check-mark boxes	
Data entry boxes	
Sliders	
VCR controls	

We hope that this exercise has been useful in introducing you to *RedShift*. On a separate piece of paper, write a sentence or two describing what you liked best about the things you did with *RedShift* today. If you are still confused about any particular button or had difficulty in getting anything to work, please make a note of that as well.

Notes

"Landmarks" in the Sky

Introduction and Purpose

In the last exercise, we introduced you to *RedShift's* many and varied functions. Today, we will begin to build on that introduction by showing you how to use *RedShift* to find your way around that "big ol' sky" out there.

You see, the sky is a little like the surface of the Earth. It has its own "geography," and like the Earth, much of the sky's geography is just plain *plain*. If you decided to begin your study of Earth's geography by looking at Kansas or the Sahara desert, for instance, you would be quickly convinced that the geography of Earth is pretty boring! The same thing can be said for many parts of the sky. However, just as there are awesome places on the Earth, so there are awesome places in the sky. The trick is knowing how to find them.

That is where *RedShift* becomes an invaluable aid. It is first and foremost your own personal "topographic" map of the sky. It will tell you not only where to find the beautiful and exciting things you've seen in pictures, but also when to look for them and under what circumstances your viewing will be the best. And, just like a good map of the Earth, *RedShift* will allow you to discover some "off-the-beaten-path" places, too.

Just as you learn your way around a new locale on Earth by memorizing landmarks, so you can learn your way around the sky by memorizing celestial "landmarks." On the Earth, typical landmarks might be distinctive buildings, broken-down fences, odd trees, or interesting rock formations. In the sky, the "landmarks" are unique groupings of stars. The largest and most easily recognized of such groupings are called constellations. But as you begin exploring the sky with binoculars or telescopes, you will find that there are odd and memorable groupings of stars on much smaller scales as well.

Constellations serve modern astronomers in another way, too. They provide some celestial "geography" to give objects of interest easily located addresses in the sky. In a very real sense, the constellations are like the states in the United States or the countries on the globe. Each has well-defined boundaries that provide address information for the objects that lie within them.

2

The purpose of today's exercise is to learn to use *RedShift's* navigation tools to find your way around the sky as well as to locate a few stars, constellations and other interesting objects. You will learn how to find the constellation boundaries using *RedShift*. You will also learn how to recognize some of the more familiar constellations, and how to use them to find other, less familiar places in the sky. First, however, we will further customize your sky display to suit your particular tastes.

Procedure
Section 2.1

Start *RedShift College Edition,* and then bring the *RedShift 3* screen to the front. (Review the instructions in Appendix A if you have forgotten how.) Before we go any farther, we need to set the parameters of the sky display with the features you will most often be using.

First, open the Display menu and click on the item labeled **Natural Sky Color** to turn the blue sky background **Off** during daylight hours. Now you should be able to see the stars on the *RedShift* screen even during the daytime. Again from the Display menu, click on the **Milky Way** item to remove its check mark and turn the Milky Way overlay **Off**.

Finally, open the Display menu and click on the **Constellations** item if it does not have a check mark. This will turn **On** the constellation pattern overlays.

Now check the panels shown on the screen. The following panels should be open: **Time**, **Control Time**, **Location**, and **Aim**.

If any of these panels is not open, open it by clicking on the appropriate small icon button in the **Settings** and **Activities** panels. For the moment, open the Control Panel by clicking on the small Control Panel icon in the Activities Panel.

When the Control Panel opens, click on the number **10** in the **Step** box and type in **5** to reduce the number of degrees by which the screen moves every time a directional arrow is pressed. Now click the small Control Panel icon once more to close the Control Panel.

Now click on the small arrow to the right of the Aim box in the Aim panel. From the menu that drops down, select **Stars**, and then click on **Polaris** as shown. The view will rotate to put Polaris, the North Star, in the center of your screen.

Section 2.2

Now we are ready to do a little customizing of your sky display. Each person has a different idea about what makes the sky easiest to use, and *RedShift* allows you to set many of the display parameters to your own liking. Click on the panel button labeled **Filters**, and from the menu that drops down, select **Guides**. In the **Local** section of the Guides filters window, click on the small **Horizon** box, and select **Line** from the drop-down menu. Now click on the small **Stars** tab at the left of the screen to switch to the Stars filters. Selecting any of the tabs on the left side of this window will change the Filters window to that particular type of object.

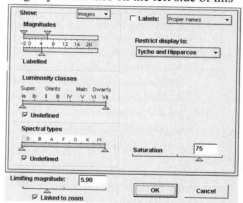

In the Stars filters window are several slider controls. We will set the sky for a typical dark sky site high in the mountains or way out on the desert. Set the upper **Magnitudes** slider and the **Limiting magnitude** slider as shown. Change the entry in the data selector **Restrict display to:** so that it reads **Tycho and Hipparcos** as shown.

As you make each change, notice how the small **Show Preview** display changes. This small display allows you to preview any changes you may wish to make before resetting the entire sky. When you have made all these changes, click **OK** to apply the changes to the entire sky.

Now look at the stars of the Little Dipper (Ursa Minor). You will see that all the stars are now brighter and that there are more stars. If this view seems too artificial to you, you may wish to re-open the Stars Filter window and tinker with these adjustments. The upper **Magnitudes** sliders

2

allow you to set the brightest and dimmest stars to show. The upper left-hand slider determines the magnitude of the brightest stars displayed. Move it to **5**, and look at the Preview display. You have just eliminated all stars brighter than 5th magnitude—those easiest to see. Restore the left-hand slider to its maximum position.

Now slide the upper right-hand slider all the way to the right. This tells the computer to display even the very faintest stars in its catalogue. You will not immediately see the effects of this setting, however, because of the **Limiting magnitude** slider. The **Limiting magnitude** slider lets you determine the sensitivity of your view. The sensitivity of the typical human eye in a very dark sky is about 6th magnitude, so we suggest you set it to that value. However, if you wish to see where all the fainter stars are, drag the **Limiting magnitude** slider all the way to the right and look at the Preview display. You should see so many stars that it will be hard to see any blank sky. For maximum usefulness, leave the upper right-hand **Magnitudes** slider at its maximum position and vary the sensitivity of the view with the **Limiting magnitude** slider.

There is one last option to note here: the **Linked to zoom** box. When this option is checked, the sensitivity of the view screen increases as the telescope magnification is increased, just as a real telescope would do. This allows you to see a normal sky at a zoom of 1 but still see the very faintest stars when the zoom is set to a maximum value. We will almost always leave this option checked.

When you think you understand how each control works, set the sliders to create a sky most like what you are used to looking at. We recommend that you set the upper **Magnitudes** sliders to their extreme positions and then vary the view from a city view to a dark sky view by adjusting the **Limiting magnitude** slider. Click **OK** to see the effects of your settings. You may have to reset the **Limiting magnitude** slider several times to get a sky that looks like the one you're familiar with.

Once you have the **Limiting magnitude** set as you like, again open the Stars Filter window and turn to the **Saturation** slider. Move it to its maximum position and click **OK**. What happens to the colors of the stars? Move it all the way to the minimum position and click **OK**. Now what are the colors? The ability to show star color makes *RedShift* a powerful tool for identifying different types of stars, but it can be confusing because our eyes do not detect any but the brightest star colors. Try several different settings to get the **Saturation** slider set to a level you find easiest to use.

Section 2.3

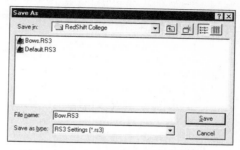

Now that you have the sky looking the way you want it, you need to save these settings. From the File menu, select **Save Settings As...** to open the **Save As** dialog box as shown.

A. Your Own Computer

If you are working on your own computer, click in the box titled **File name** and type in a unique name (preferably eight letters or less) that you will be able to remember later. Use your initials if you like, but be sure to leave the extension **.RS3**. Your completed file name should now be something like this: **Bow.RS3**. When you have entered an appropriate file name, click on the **Save** button to save your customized personal settings.

B. Classroom Computer

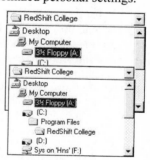

If you are using a classroom computer, place a blank floppy diskette in the computer's disk drive, and then click on the small arrow to the right of the *RedShift College* folder name (as shown). From the list of folders and drives that will appear, select **3½ Floppy (A:).** (Use the vertical scroll bar if necessary to locate drive **A:**.) At this point, the floppy drive should activate and load a blank directory.

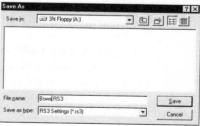

Now, click in the box titled **File name**, and type in a unique name (preferably eight letters or less) that you will be able to remember later. Use your initials if you like, but be sure to leave the extension **.RS3**. Your completed file name should now be something like this: **Bow.RS3**. Finally, click on the **Save** button to save your customized personal settings.

25

2

Section 2.4

We are now ready to use *RedShift* to find some of those awesome spots in the sky. First, turn off the **Guides** item in the Display menu by clicking on it to remove its check mark. This action will remove the horizon line at the bottom of the screen and allow you to see all of the sky. Next, click on the Filters button, and select **Stars** to open the Stars filters window. Set the **Limiting magnitude** to

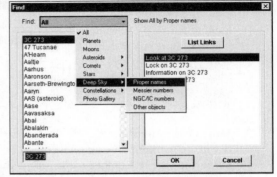

6.0 to simulate a really perfect desert sky, and click **OK**. Now it's time to learn about one of *RedShift's* most powerful tools—the Find window.

There are several ways to open the Find window. First, you can select **Find** from the Information menu at the top of the screen. Second, you can hold down the **CTRL** key and press the **F** key on your keyboard. Finally, you can click on the small arrow at the right of the Aim box and choose **Find** from the drop-down menu.

Once the Find window is open (as shown above), you can locate any of the objects in the *RedShift* database, from stars to constellations to deep-sky objects. To begin with, we will use this window to find a beautiful deep-sky object—a spiral galaxy known as the Triangulum Galaxy.

To find this galaxy, first click on the small arrow at the right edge of the **Find:** box at the top of the window. When the menu drops down, choose **Deep Sky**, and then select **Proper names** as shown here to put the Deep Sky names in the left-hand list box. When the Deep Sky list appears, scroll nearly to the bottom of the list, and double-click on the entry **Triangulum Galaxy**. Finally, click the **OK** button to rotate the screen to center on the Triangulum galaxy. When the screen rotates, there will not seem to be anything in the center. The reason for this is simple. The Triangulum galaxy is too dim to be seen without binoculars or a telescope.

To see what this galaxy would look like through a small telescope, click on the small arrow to the right of the Zoom box (as shown), and select **3.5** from the menu that drops down. Sketch the shape of this galaxy in the space below.

```
AIM
        chi Persei
Zoom 1.00    ▼
    9999
    500
    50
    10
    3.5
  Normal 1
    0.5
    0.26
  Previous  ▶
```

Now click on the Zoom box arrow, and select **50**. This is what the Triangulum galaxy would look like through a large telescope. In the space below, sketch this beautiful galaxy.

When you have finished sketching the Triangulum galaxy, again click on the Zoom box arrow, and select **Normal 1** to return the magnification to 1. Now do the following exercises.

Q1 Write down the names of all the constellations that you can see on the screen.

Q2 Look carefully at the sky display on your screen. The Triangulum Galaxy is not actually inside of any constellation outline. It could actually be called by any one of three constellation names. What are they?

Q3 Now click on some stars that are between the constellation pattern of Triangulum and the other surrounding constellation patterns. Write down the name abbreviations of any five such stars in the space below.

Now let's interpret those abbreviations. "BSC" stands for Binary Star Catalogue; "And" stands for Andromeda. "Psc" stands for Pisces, and "Tri" stands for Triangulum.

Q4 Click on the star at the apex of the Triangulum outline, and then click on its name tag to open its information box. Write down its proper name. Just to the right of the **Object:** box, you will find the abbreviations for its Flamsteed and Bayer names. Write down these abbreviated names.

The complete Bayer name of this star is **Alpha Triangulum**. The complete Flamsteed name of this star is **2 Triangulum**.

2

Section 2.5

By now, you should be getting the idea of how most of the stars and deep sky objects are named. They are most often identified by their constellation addresses. Aside from the "Proper" names of the very brightest stars (which come down to us from the Arabic translation of Ptolemy's *Almagest*), we refer to most stars by their Bayer designations, such as "Alpha Centauri" or "Gamma Leonis." The "Alpha/Beta/Gamma" designations are given according to the brightness of the stars in each constellation. Alpha denotes the brightest star in a constellation (at least as measured by Bayer), Beta corresponds to the second brightest, etc.

Most of the bright stars and beautiful deep-sky objects lie within the constellation outlines. However, many others such as M33, the Triangulum Galaxy, do not. So astronomers came up with a simple way to give these "in-betweeners" constellation addresses anyway. They arbitrarily defined constellation boundaries, just as cities, states, and countries have done here on Earth. Then any star, planet, asteroid, comet, or deep-sky object found within the constellation's boundary is said to be "in" that constellation.

RedShift allows us to see just where those boundaries fall. Click on the **Filters** button in the control panel, and select **Constellations** from the drop-down menu as shown here.

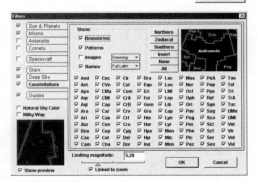

In the Constellation filters window that opens, click in the box titled **Boundaries** (as shown here) to display the constellation boundaries on the screen. Click **OK**, and then answer the following questions.

Q5 With the boundaries turned on you can see the official edges of each constellation. What constellation is M33, the Triangulum Galaxy, *almost* in? (You may have to click the <Down> arrow on your computer's keyboard to find its name.)

Q6 Open the Find window (either from the Information menu or by holding down the **CTRL** key and pressing **F**). Click the small arrow at the right of the **Find:** box, choose **Constellations**, and select **Latin** (as shown) to put the familiar constellation names in the left-hand list box.

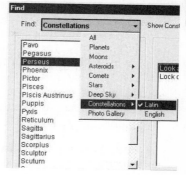

In the list box, scroll down to **Perseus**, double-click on the name, and then click the **OK** button to rotate the view to center on the constellation of Perseus.

Use the directional arrows on your computer's keyboard to adjust your view right and left (or up and down) until you can also see the constellation of Cassiopeia. Between Perseus and Cassiopeia, you will find what looks like a pair of twin stars. Click on this pair, and you will find that it is really a pair of open star clusters. Write down the names of these two open clusters.

Q7 These two star clusters (Q6) are called the Perseus Double Cluster and can be seen easily with binoculars. If the constellation boundaries had been drawn a little differently, what might they have been called? (What constellation are they *almost* in?)

Position the cursor right between the two, and double-click the left mouse button. The screen will rotate to center on that spot. Now click in the Zoom box, type **20**, and press the <Enter> key on the computer keyboard to see what these clusters look like in a telescope. When you have had a good look at this beautiful pair of star cities, change the Zoom back to **1**.

Section 2.6

Now let's use your knowledge of *RedShift's* Find window to locate several beautiful galaxies, nebulae, and star clusters. Select **Find** from the Information menu to open the Find window. In the Find window, choose **Deep Sky** in the **Find:** box and select **Proper Names**. Scroll through the list box to find each of the deep-sky objects in the data table below. When you find the object, double-click on it, and then click the **OK** button to center the screen on that object. Since most of these objects will not be visible with a magnification of 1, change the zoom factor to **10** to make

2

sure the object is really in the screen's center, and then change it back to
1.

Now note the constellation boundaries to discover which constellation the
deep-sky object is officially in. (You may have to move the display
around with the keyboard arrow keys to reveal the constellation name.)
Write this name in the Constellation column of the data table, and repeat
the process until you have found all the objects in the table. Each of these
objects is beautiful or interesting. To see photos of them, click on the
Information menu, and select **Photo Gallery**.

To see photos of the Sombrero and Whirlpool galaxies, select **The
extragalactic universe** from the Contents list, and click on **Spiral
galaxies** to open the Spiral galaxies photo list. Scroll through this list to
find photos of these two beautiful galaxies. When you locate each photo,
click on the **Magnify** icon  on the thumbnail image to enlarge the
picture to full-screen size. To return to the photo list, click on the
Thumbnail icon .
Centaurus A is a radio galaxy. To find its photo, click on **Peculiar and
active galaxies** in the **Extragalactic universe** section, and then select
Radio galaxies.

The Ring nebula and the Dumbbell nebula are known as *planetary
nebulae*. To see photos of these, select **The Galaxy**, and click on
Planetary nebulae to open this photo list. Scroll down through the list to
locate each one. As before, use the Magnify button to enlarge each image
and the Thumbnail button to return to the photo list.

You can find photos of the Lagoon and Trifid nebulas in the **Interstellar
matter** page, the Crab nebula in the **Supernova remnants** page, and the
Pleiades and the Jewel Box cluster in the **Open clusters** page, all in **The
Galaxy** section. When you are finished looking at the pictures, click on
the **Close** button at the top right corner of the Navigation bar to return to
RedShift 3. (Be careful not to click the Close button for the entire
RedShift window!) *(Macintosh users will click the small square in the
upper left corner of the window to close it.)*

Data Table

Constellation Addresses of some Deep-Sky Objects			
Object Name	Constellation	Object Name	Constellation
Black-Eye Galaxy		The Pleiades	
Crab Nebula		Ring Nebula	
Dumbbell Nebula		Sombrero Galaxy	
Jewel Box		Trifid Nebula	
Lagoon Nebula		Whirlpool Galaxy	

Section 2.7

To conclude today's exercise, we will show you a neat way to find these galaxies, nebulae, and star clusters using the constellations. None of these deep-sky objects are visible from typical suburban neighborhoods (except the Pleiades), but you can see many of them in binoculars or telescopes.

Because these objects are not visible to the unaided eye, you will need a way to locate them. One of the best techniques for this is called star-hopping. It will not only help you find deep-sky objects, but it will get you better acquainted with the stars at the same time. The idea behind star-hopping is to start with a constellation we can find (like the Big Dipper or Orion) and then plot a path from star to star until we arrive at the area we wish to observe. Then, with binoculars or a telescope, we can zoom in on that spot and usually find our targets without too much trouble.

In the following section, we will plot a star-hop to locate the Andromeda Galaxy, M31. We will begin at the big "W" of Cassiopeia and work our way from star to star till we arrive at M31. Before we begin today's star-hop, however, we need to change some of *RedShift's* settings.

First, click on the Filters button in the control panel, and select **Constellations** to open the Constellations filters window. Click on the **Boundaries** box to remove the check mark, and turn off the constellation boundaries. Now click on the left-hand tab labeled **Deep sky** (the tab, not the box) to change the Filters window from Constellation settings to Deep Sky settings. We are going to make several changes to the Deep Sky settings.

2

In the upper right section of the Deep Sky window, click in the **Labels** box to turn the labels on in the sky display so that you can see the names of the various deep-sky objects. (We usually leave these names off because they can easily clutter the sky.) Then click on the small button to the right of the Labels box (initially titled **Proper names**), and select **Messier numbers** from the menu that drops down. With this setting, you will see only the numbers of the objects instead of lots of long names.

Next, locate the **Show:** slider in the upper left part of the Deep Sky window. Click on the lower slider marked **Labelled**, and drag it to **10** on the slider scale to set the magnitude of the faintest labeled objects at 10. Now skip down to the **Galaxies** section in the lower left part of the Deep Sky window. Click on the small button initially titled **Images**, and select **Icons** from the drop-down menu so you can see the positions of galaxies even when the galaxies themselves are too small or dim to see. When the Deep Sky Filters window looks like the window below, click **OK** to make all these changes.

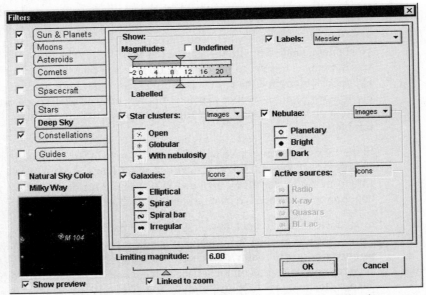

Now that the sky settings are ready, you will also need to set the time, date, and zoom settings as shown on the next page.

Date	06/01/1999
Time	10:00 p.m.
Zoom	0.5

The date and time are set by clicking in the Time settings panel (shown here) and typing in the new date and time. And, of course, the zoom box is at the bottom of the Aim panel.

TIME Local ▼
08 / 30 / 1999 AD ▼
-5.0 h 10 : 09 am ▼

Section 2.8

When you are all done, you should be able to see the constellation patterns and names (without the boundary lines), and you should see lots of small orange icons with M labels marking the positions and names of galaxies.

Now that the sky, date, and time are set, let's rotate the screen to center on the part of the sky we are interested in. Click on the small arrow to the right of the Aim box, select **Deep Sky**, and click on **Andromeda Galaxy** to center on our goal.

You should now see the constellations of Cassiopeia, Andromeda, Pegasus, Perseus, and Triangulum all on the screen. Our quarry for this exercise, M31, should be visible as a small galaxy icon right in the middle of the constellation of Andromeda.

We are now ready to begin. The first leg of the star-hop follows the inside diagonal of the big "W" of Cassiopeia as shown. Trace a line down the inside of the "W" and extend it until you come to a bright star that forms one corner of the big square of Pegasus.

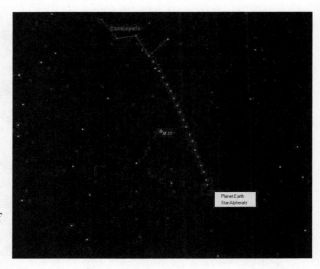

This star is called Alpheratz. This is a long hop, but it is the only bright star in line with the inside leg of the "W." Click on the star and check its name tag to make sure it is Alpheratz. Now increase the Zoom to **1** so that you can see the star-field more realistically.

Now hop at almost a 90 degree angle to the left (west) as shown here. Go past the first bright star, and then continue about the same distance again, veering slightly to the north, and you will see an even brighter star. This second star, Mirach, is a pretty yellow color.

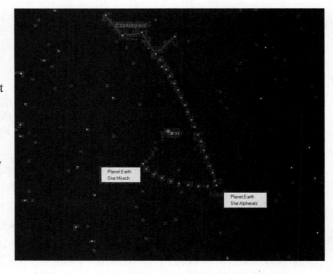

Again, click on the star to see its label to make sure you have found Mirach.

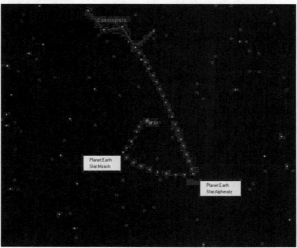

From Mirach, hop northward through two faint stars to the star Nu Andromeda.

M31, the Andromeda Galaxy, will be found by sweeping your binoculars just one full field to the east from this faint star.

Section 2.9

There are lots of other objects like the Andromeda Galaxy that you can see easily with binoculars if the sky at your observing site is really dark. Your star-hop challenge for the remainder of this exercise will be to chart your own course to one of these deep-sky objects. If you are doing this exercise in the fall, chart a star-hop path to H and Chi Persei, the Perseus double cluster we looked at earlier in this exercise. If you are doing it in the spring, chart a star-hop path to M44, the Beehive Cluster. Step-by-step instructions to get you started are given below.

1) First, set the Zoom to an ordinary wide-angle view of the sky—**0.5** or **0.6**—and press the keyboard <Enter> key. Click on the Filters button and select Deep sky to open the Deep sky Filters window. Set the **Labelled** object magnitudes to **8** and the Deep sky labels to **Proper names**. Then change the Limiting magnitude to **5.1** to simulate a typical suburban sky.

   Now click on the Constellations tab. Click on the small Names box and change it to **Abbr. Latin** to change the constellation names to abbreviations to keep them from covering up the object names. When you are all done, click **OK**.

2) Once the display parameters are set, you need to locate the object you wish to find. Open the Find window, select **Deep Sky**, and click on **Proper names**. Scroll down and double-click on the object you are looking for (either **H and Chi Persei**, or the **Beehive Cluster**), and click **OK**.

3) Find a prominent constellation from which to begin. This constellation should be fairly close to the object you wish to find and should be one you will be able to identify and find in the real sky. For H and Chi Persei in the fall, you should begin from Cassiopeia or Pegasus. For the Beehive Cluster in the spring, you should start from Leo or Orion.

   a) To start from Pegasus (Peg) to find H and Chi Persei, locate the large square of stars that forms the body of Pegasus (Peg). The blue-white star at the corner of the square nearest H and Chi Persei is Alpheratz. Alpheratz will be your beginning point for this star-hop. Use the directional arrows to move the sky until the screen is centered just about halfway between Alpheratz and H and Chi Persei.

   b) To start from Leo to find the Beehive Cluster, find the backward question mark that forms the head and mane of Leo. (It also looks like a sickle or like the curved top of a coat-hanger.) The "sickle" will be your jumping-off point for this star-hop. Use the directional arrows to move the sky until it is centered about halfway between the "sickle" and the Beehive Cluster.

4) Now increase the Zoom factor, keeping both your jumping-off point and your object on the screen. (You may have to move the control panels to see one or the other.) If both are not visible, use the directional arrows to re-center the screen. The "Great Square of Pegasus" and the "Sickle" of Leo's head and mane are both fairly easily found and identified in the real sky, so these are both realistic starting points for finding H and Chi Persei or the Beehive Cluster. You are now ready to start plotting your star-hop.

5) Next, make a printout of the sky display with your object and the jumping-off constellation on it. Here you have a choice: You can leave the constellation patterns on your printout or remove them. If you wish to remove the patterns, open the Constellations filter window, click in the **Patterns** box to remove the check mark, and click **OK**. Now choose **Print** from the File menu to print your star chart.

6) Once you have a printout of the deep-sky object and your jumping-off-point constellation, work out a star-hop path to your object. Look for distinctive groupings of stars like straight lines, triangles, double stars, etc., around or between the constellation and the deep-sky object. Sometimes you may even find groupings of stars that actually seem to point toward the object you're looking for. In general, don't choose star-hops that are very far apart, or they may be difficult to follow in the real sky.

As you go along you'll find that star-hopping is fun. It's also a very personal thing. Our students have seen everything from unicorns to fox terriers to Eddie Van Halen's guitar in the stars!

7) On your printout, circle each distinctive grouping of stars (straight line, triangle, double star, etc.), and draw an arrow from each one to the next nearby distinctive group, and then to the next, until you arrive at your deep-sky object. You may take several star-hops to arrive at your destination or only one or two.

When you have finished marking your star-hop on the printout, staple it to the back of this exercise.

Discussion

Constellations, as you have seen in this exercise, are much more than just crazy stories in the stars. They give us familiar "landmarks" to help us find our way around the sky. They also provide a way to break the sky up into well-defined, manageable areas for locating other objects like star clusters, galaxies, nebulae, and planets. Finally, constellations, and smaller distinctive star groupings can ultimately help us plot paths through the sky to places that would otherwise be very difficult to find. With a little practice, we can chart paths to lead us to just about any destination we choose.

Conclusion

In a couple of sentences, summarize the uses of constellations. Tell what you have learned about specific constellations in today's exercise, and give your thoughts about the process of star-hopping.

Optional Observing Project

Do the observation exercise "Constellations" in Appendix C.

2

Notes

Finding Your Way Around (in the Dark!)

Introduction and Purpose

When would-be astronomers get their first telescope, they can hardly wait to set it up so that they can at last see with their own eyes those wonderful celestial sights they've been reading about. Unfortunately, that first encounter with the real sky often becomes the last, as many future astronomers decide to leave the field at that point. Their problem is one that perhaps you have also encountered—when you start scanning the sky with a high-magnification telescope, you find that there is a *lot* of dark sky out there with (apparently) not very many interesting things in it.

Another frustrating aspect of stargazing is trying to tell someone else how to find the object you are looking at. Phrases like "right above that bright star" or "just above that tree" don't mean much when the horizon is ringed with trees and the sky is full of bright stars.

In the last exercise, you learned how to use constellations to get a general idea of where to find planets, deep-sky objects, and stars. In today's exercise, we will develop a more precise method for locating star positions at a given time from a given place on Earth. You will learn how to express the position of a star, planet, or satellite in terms that everyone can understand, and you will learn more about navigating through the *RedShift* sky as well.

Procedure
Section 3.1

Start *RedShift College Edition*, and then bring the *RedShift 3* screen to the front. When the *RedShift 3* screen opens, click on the File menu and select **Open Settings** to open the personal settings file you saved in the last exercise. (Review Appendix B if you have forgotten how to do this.)

Section 3.2

When you reopen your settings, the time and date are set as they were when you last saved the settings file. What you see in the sky depends not only on where you are, but on what time it is, so we need to reset these important parameters.

Let's reset the time to a good observing time, say, 8:30 p.m. The Time settings panel should already be open. Click on the **hours** part of the time, and the background color of the number should change. Type in the hour for which you wish to set the sky, **8**. Then press the <Tab> key on your computer keyboard or click on the **minutes** part of the time, and type in the rest of the value, **30**. If the **am/pm** part of the time is set to **am**, then press the <Tab> key again, and type in **pm** (or click on the small arrow at the right of the box, select **am/pm**, and click on **pm**). Now the time box should read **8:30 pm**. Change the **Date** to today's date in the same way.

Section 3.3

Now that you have the date and time set, let's find our directions. Click in the Zoom box, type in **0.8**, and press the <Enter> key on the keyboard. If you cannot see the green horizon line (or if the whole sky is green), click on the <Up> or the <Down> arrow button on the computer keyboard until you see the green horizon line going across the screen. You should now be able to see a green **N** (for "north") near the center of the horizon.

Section 3.4

Now click on the Filters button, and select **Guides** to open the Guides filters window. In the **Local (Azm/Alt)** section of this window, click on the small box to the right of the **Horizon** entry, and select **Transparent** from the drop-down menu. Now click on the **Celestial (RA/DEC)** box to open that section. Click on the **Grid** box to remove that check mark, and

Local (Azm/Alt):		Celestial (RA/Dec):
☐ Zenith/Nadir		☑ Poles
☑ Horizon [Transparent ▼]		☐ Equator
☐ Meridian		☐ Meridian
☑ Compass		☐ Vernal equinox
☐ Grid		☐ Grid
☐ Numbering		☐ Numbering

then click on the **Poles** box to put a check there. When the top part of the window is as shown here, click **OK** to make these changes.

Above the horizon, you should now see a blue marker labeled **NCP**. This abbreviation stands for "north celestial pole." It is a point in the sky directly above the Earth's North Pole.

You are now seeing the sky as it would look tonight at 8:30 p.m. if you were facing north. Look at the center of the status bar at the bottom of the screen. There are two sets of numbers here labeled **Azm:** and **Alt:**. These stand for "*azimuth*" and "*altitude*." These are direction angles based on your local horizon. Note that both altitude and azimuth measurements have two parts. The first part of the number is measured in *degrees* (°). The second part of the number is measured in *arc-minutes* ('). An *arc-*

minute is 1/60 of a degree (just as a minute is 1/60 of an hour). Since 30 minutes is ½ hour, so 30 arc-minutes is ½ a degree. Thus, **30° 30'** is equivalent to **30 1/2** degrees. While looking at the status bar, answer the following questions.

Q1 Move the mouse right and left across the sky, and watch the Alt/Azm values in the status bar. Which are you changing by moving parallel with the horizon, the altitude or the azimuth?

Q2 Now move the mouse up and down on your screen and watch the Alt/Azm values in the status bar. Which measurement is perpendicular to the horizon, the altitude or the azimuth?

Q3 Move the mouse so that the cursor is positioned directly over the **NCP** marker. What is the azimuth of the Earth's celestial North Pole (the azimuth of true north)? *Write only the number of degrees (rounded off).*

Q4 What is the altitude of the Earth's celestial North Pole as viewed from your viewing location? *Write only the number of degrees (rounded off).*

Q5 Now, look at the Location Panel. What is the latitude (Lat) of your viewing location? *Write only the number of degrees (rounded off).*

Q6 Comparing Q4 and Q5, what might you guess about the relationship between an observer's latitude and the altitude of the celestial North Pole above the horizon?

Q7 There is a star very near the NCP. You can find your latitude anywhere in the Northern Hemisphere simply by measuring the altitude of this star. Click on it to reveal its name tag and write its name here.

Q8 Move the cursor back and forth slightly to the left of the celestial North Pole. What is the maximum value an azimuth measurement can have? *Write only the number of degrees (rounded off).*

Q9 Now move your cursor down to the horizon. What is the altitude measurement? *Write only the number of degrees (rounded off).*

Q10 Now position the cursor below the horizon. Notice the altitude value. How can you tell just by looking at a star's altitude whether it is below the horizon?

'.e a half of a sphere sitting above our heads, with one edge of
e on one horizon and the other edge on the other horizon.
...e horizon up to the top of the sky (the *zenith*) and back
. .o the other horizon is 180 degrees (half a circle). So how many
degrees in altitude would it be from the horizon up to the *zenith*, the very
top of the sky?

Section 3.5

Now let's move around in the sky a little. Press any of the directional
arrows on your keyboard, and watch what happens to the sky display.

First let's find the Big Dipper (Ursa Major). If you are an explorer, go
ahead and use the directional arrows and move the sky around till you can
see all seven of the main stars in Ursa Major, the Big Dipper. If you want
to go there directly, open the Find window (either from the Information
menu or from the keyboard). In the Find window, click on the find arrow,
select **Constellations**, and click **Latin** to put the familiar constellation
names in the find window. Now scroll down through the constellation list,
double-click **Ursa Major**, and click on the **OK** button. The view screen
will move around until Ursa Major is in the center of the screen. Now do
the following.

Q12   Position your cursor over the end star in the Big Dipper's cup, Dubhe.
Click on the star to verify its name. From the status bar, find Dubhe's:
        altitude (   ) and
        azimuth (   )
from your location at 8:30 p.m. tonight. *Write only the number of degrees
(rounded off).*

Q13   If you wanted to find the Big Dipper tonight at 8:30, from your location
would you look:
        A)   A few degrees east of north (how many degrees)?
        B)   A few degrees west of north (how many degrees)?
        C)   Straight north?
        D)   In some other direction (in which direction)?

Section 3.6

Q14 Use the directional arrows to move the sky to the right till you can see the green **NE** (northeast) marker on the horizon. Position your cursor over the green **NE**. From the status bar, read the azimuth corresponding to a direction of due northeast *(degrees only)*, and write it here.

Q15 Use the directional arrows to continue moving the sky to the right till you can see the green **E** (east) marker on the horizon. Position your cursor over the green **E**. What is the azimuth corresponding to a direction of due east *(degrees only)*?

Q16 What is the difference between the azimuths of northeast (Q14) and east (Q15)? **Q14 - Q15** = (   )

Q17 Add that difference (Q16) to the azimuth of east (Q15) to predict the azimuth of southeast. **Q16 + Q15** = (   )

Move the screen around to the **SE** marker and check your answer.

Q18 Use the directional arrows to continue moving the sky to the right till you can see the green **S** (south) marker on the horizon. Position your cursor over the green **S**. What is the azimuth corresponding to a direction of due south *(degrees only)*?

Q19 Note the difference between the azimuths of east (Q15) and south (Q18). On the basis of that difference predict the azimuth of west. Now move the screen around to the **W** marker and check your answer.

Section 3.7

To understand what is going on, imagine you are standing in the center of a large circle that is divided into 360 equal segments. If you divide 360 by 4, you will understand why the distance between each cardinal point is 90 degrees. Study the accompanying diagram, and fill in the missing numbers and directional points.

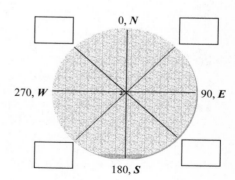

Q20 If someone spotted a satellite while out star-gazing and told you it was at an azimuth of 130 degrees, would you look:
A)   Approximately southeast?
B)   Approximately south?
C)   Approximately southwest?
D)   None of the above

Q21 If you spotted an odd star in the northwestern sky and wanted to describe its location to someone else, what would you give as its azimuth?

Section 3.8

Once again, open the Information menu. You will see the notation "Ctrl + F" beside the **Find** entry. As we mentioned before, this is a shortcut to open the Find window. Instead of selecting **Find**, close the menu by clicking anywhere else in the sky. Then hold down the <Ctrl> key and press the <F> key simultaneously to open the Find window. In the Find window, scroll down through the constellation list, this time double-clicking on **Scorpius**, and click **OK**. Scorpius should appear in the center of the screen. Now answer the following questions.

Q22 Is Scorpius above the horizon at your location tonight at 8:30? (Anything below the horizon has a green background.)

Q23 One end of Scorpius is shaped like a fishhook—this is the tail of the scorpion curled over his back. Click on the last star in the scorpion's tail (the end of the sharp curve). What is its name?

Q24 What is the azimuth of this star as seen from your location at 8:30 tonight *(degrees only)*?

Q25 What is the altitude of this star as seen from your location at 8:30 tonight *(degrees only)*?

Q26 If someone spotted a satellite while out star-gazing and told you it was at an altitude of 85 degrees, would you look:
A)   Near the horizon?
B)   Near the zenith?
C)   Halfway up the sky?
D)   None of the above

Q27 If you spotted an odd star halfway up to the zenith and wanted to describe its location, what would you give as its altitude?

Q28 Change the time in the Time settings panel to **9:30 pm** and press the <Enter> key on the computer keyboard. Now, check the altitude and azimuth of the tail of Scorpius. Does a star or constellation always have the same altitude and azimuth?

Section 3.9

Open the Find window (either from the Information menu or the keyboard), click in the **Find:** box, select **Stars** from the drop-down menu, and click on **Proper names.**

In the data table below are 5 **Proper** star names and 5 **Bayer** star names. We will begin with the 5 proper star names. Scroll down through the star list, and double-click on the first proper star name listed in the data table below.

When you have spotted the appropriate star name, click **OK** to move the view screen around to center the star. Now click on the star in the center of the screen to verify it is the correct star. Once you open the star's name tag, look in the status bar to find the star's altitude and azimuth, and record them in the data table. Now, repeat this process of opening the Find window (use <Ctrl> + <F> to do this faster) and finding and recording the data for each of the stars in the first column.

Re-open the Find window once again. Again, click on the **Find:** box and select **Stars**, but this time click on **Bayer/Flamsteed**. You will see the list of stars change to the Bayer/Flamsteed designations. Now repeat the steps listed above to find the altitudes and azimuths of the five stars in the data table.

Data Table

Altitude and Azimuth					
Proper Star Names	Altitude	Azimuth	Bayer Star Names	Altitude	Azimuth
Achernar			Alpha Auriga		
Albireo			Alpha Bootis		
Canopus			Beta Andromeda		
Denebola			Gam_1 (Gamma 1) Leonis		
Zuben Elgenubi			Eps_1(Epsilon 1) Lyrae		

3

Discussion

In this exercise, you have learned more about using two of *RedShift's* powerful navigational tools—the Find window and the keyboard directional arrows. You learned that you can open some of *RedShift's* tools either from the menus or by using keyboard shortcuts. For tools that are used a lot, like the Find tool, keyboard shortcuts can be a great time-saver.

You learned how to select different catalogues for the star names and how to get positional information from the sky display. You also learned how to place compass point markers along the horizon and how to interpret the altitude and azimuth coordinates for stars and constellations.

You learned that altitude and azimuth are the measurements we use to describe the location of celestial objects with respect to our viewing location. They tell us just where to look with respect to the horizon and the points of the compass. However, these coordinates change from hour to hour as the Earth turns and the stars and constellations move across the sky. As a result, altitude and azimuth coordinates are useful for only a few minutes.

In order to provide location coordinates that do not change every few minutes, astronomers have come up with another celestial coordinate system called Right Ascension and Declination. In the next exercise, we will explore this useful location system, as well as continue our work with the navigation tools of *RedShift*.

Conclusion

On a separate sheet of paper, summarize the altitude/azimuth coordinate system. Include the units of altitude and azimuth, and mention the values of altitude for the horizon and the zenith as well as the azimuth values for the four cardinal compass points.

Address ~~Un~~*Known!*

Introduction and Purpose

In the last exercise, we explored the local observer system of locating celestial objects known as altitude-azimuth (Alt/Azm for short). We discovered, however, that in this system, the location coordinates of any star or constellation changes from minute to minute as the Earth turns. It is impossible, therefore, for you to tell someone else on the other side of the world (or even of the state) where to find something in the sky—or even to find it yourself a few hours later.

Astronomers quickly realized the need for an absolute celestial coordinate system—one that wouldn't change as time went by, and that would be the same for people all over the world. The solution to this problem is a universal celestial coordinate system called ***Right Ascension-Declination*** (RA/Dec for short).

The purpose of today's exercise is to find out more about the Right Ascension-Declination coordinate system. In the process, we will continue to develop our skills with *RedShift's* navigation tools. We will also discover the key to finding most of those awesome objects we see in the books and magazines on astronomy.

Procedure

Section 4.1

Start *RedShift College Edition*, and bring the *RedShift 3* screen to the front. When the *RedShift 3* screen opens, click on the File menu, and select **Open Settings** to open your personal settings file. (Review Appendix B if you have forgotten how to do this.) The only thing we will need to change is the date. Click on the date portion of the Time settings panel. Change whatever part of the date is incorrect so that it displays today's date, and press the <Enter> key on the computer keyboard.

Section 4.2

As before, you should be looking north, and the compass point **N** should be in the center of the horizon. Click on the Filters button and select **Guides** to open the Guides filters window. In the **Local (Alt/Azm)** section of the window, click on the small button to the right of the **Horizon** box, and select **Transparent** to allow you to see through the horizon. Then click on the **Zenith** box to turn that feature on. In the **Celestial (RA/Dec)** section, click on the box to activate that section. Make sure the **Grid** box in this section is checked. Now click on the **Poles** box, the **Equator** box, and the **Numbering** box to turn on those features. Finally, click **OK** to record your changes.

If the **NCP** label is not visible on the sky display, click on the <Up> or <Down> arrows on the keyboard until the label is just visible near the top of the screen. Now answer the following questions.

Q1 Look at the lines radiating out from the NCP. Follow one of the lines until it intersects with a circular line. At this point, there are two numbers. The top one is the label for the straight, radiating line, and the bottom one is the label for the circular line. What two letters are used as units for the top number?

Q2 Find the star **Polaris**. It is the closest star to the NCP. Which radial line is Polaris closest to? (It is just barely between two of the lines. They are very close together at that point. You may have to use the directional arrows to move the sky so that you can see the number labels for those radial lines.)

Q3 Position the cursor over Polaris and click on it to open its name tag. Now, click on the name tag to open an information box about the star. In the information box, you will find coordinates listed for this star. Which one of the two coordinates shown in the information box most closely matches the number on the radial line closest to Polaris (Q2)?
A)   RA
B)   Dec
C)   Alt
D)   Azm

Q4 What is the exact right ascension (RA) address of Polaris?

Q5 In the RA address, the **h** unit stands for hours. What do you think the **m** and **s** stand for?

Q6 Look at all the right ascension (radial) lines. What is the highest value these lines can have before they start over at 0? (Remember, the top numbers are the labels for the right ascension lines.)

Q7 What is the exact declination (Dec) address of Polaris?

Q8 Look at the units in the declination address. You are already familiar with the first two units in the declination coordinate from your study of the Alt/Azm coordinates in the last exercise. What does the first unit, °, stand for?

Q9 Just as with altitude and azimuth, the second unit, ', stands for arc-minutes, where one arc-minute = 1/60 degree. The third unit, ", is 1/60 arc-minute. What do you think the name of the third unit is?

Section 4.3

Press the <Up> arrow on the keyboard and wait for the screen to reset. Do this several times until the green **Z** (Zenith) marker appears on the screen. Now answer the following questions.

Q10 By looking at the grid, estimate the right ascension and declination coordinates of the zenith. (Click on a nearby star to open its name tag, and then click on the name tag to find the star's RA and Dec address to confirm your guess.)

Q11 Now click on the hour number in the Time settings panel. Increase the number by **1** and then press the <Enter> key on your computer keyboard. Now what is the approximate RA coordinate of the zenith?

Q12 The zenith is the point directly overhead. On the basis of your answer to Q11, deduce the RA coordinates of the constellation that would be directly overhead six hours later.

Change the time setting to six hours later and check your answer.

Section 4.4

Now you should understand why right ascension is measured in hours, minutes, and seconds. Each hour of the Earth's rotation moves one hour of right ascension across the sky. That is why the maximum value for right ascension lines is 23 hours. When 24 hours have passed, the cycle begins all over again, and we are back at 0 hours on the celestial sphere. Now press the <Down> arrow on the keyboard and wait for the screen to reset. Continue pressing the arrow key until you come to the line whose declination is **0** degrees. This is the *Celestial Equator*. It is an imaginary line running across the sky directly above the Earth's equator. Now answer the following questions.

Q13 If you were standing on the Earth's equator, looking at a star straight overhead, what declination would that star have?

Q14 What does it mean if the declination of a star is a negative number?

Q15 Look back at your answer to Q7. This is the declination of Polaris, the North Star. On the basis of its declination, what do you think is the highest declination value any celestial object could have?

Q16 If you were standing at the South Pole of the Earth (brrrr!), what would be the declination of an object at your zenith?

Example Problem 1: Imagine you are standing outside some evening at 8:30 p.m., looking straight up at the Zenith. If you watch long enough, you will notice that the stars are moving slowly westward. The constellation straight overhead at 8:30 has a right ascension of $6^h 25^m$ and a declination of $0°$. At 2 a.m., you awaken to find a new constellation directly overhead. What will the right ascension of this constellation be?

E1.1 To answer this question, first figure out how many whole hours it is between 8:30 p.m. and 2 a.m. (Count on your fingers if needed.) Write the answer here: **(5 hours)**

E1.2 Now convert any additional fractions of an hour into minutes. Write that answer here: **(1/2 hour = 30 minutes)**

E1.3 Now add the whole hours to 6.
$$6^h + \textbf{(E1.1)} = \text{number of hours}$$
$$6^h + \textbf{(5)} = \textbf{(11)}^h$$

E1.4 Add the minutes to 25.
$$25^m + \textbf{(E1.2)} = \text{number of minutes}$$
$$25^m + \textbf{(30)} = \textbf{(55)}^m$$

Now put them both together: $\textbf{(11)}^h \textbf{(55)}^m$
Now use the arrow keys on the keyboard to follow the Celestial Equator line around to $11^h 55^m$. You will find that this RA (with a Dec of $0°$) is at the edge of the constellation Virgo.

For questions Q17–Q20, follow the steps in the example just given. Suppose the center of Virgo (Dec $0°$, RA $13^h 0^m$) is straight overhead at 9:00 p.m.

What constellation will be straight overhead at 1:30 a.m. the next morning?

Q17 How many whole hours is it between 9:00 p.m. and 1:30 a.m.?
Write the answer here: (   )

Q18 Convert any additional fractions of an hour into minutes.
Write that answer here: (   )

Q19 Add the whole hours to Virgo's RA hour measurement, **13** ^{h}.
$$13^h + (Q17) = \text{number of hours}$$
$$13^h + (\quad) = (\quad)^h$$

Q20 Add the minutes to Virgo's RA minute measurement, **0** ^{m}.
$$0^m + (Q18) = \text{number of minutes}$$
$$0^m + (\quad) = (\quad)^m$$

Now put them both together: (   ) ^{h} (   ) ^{m}

Use the arrow keys on the keyboard to move along the Celestial Equator (keeping the Dec at 0°) until you find the position corresponding to the above RA address. In which constellation is this address?

Example Problem 2: Suppose that Orion is high in the sky at 8:00 p.m., but you want to look at something in Leo. How late would you have to stay up in order to see Leo high in the sky?

E2.1 Click on the small arrow to the right of the Aim box, select **Constellations**, and click on **Orion**. What is the approximate right ascension of the center of Orion? (Use the position of the name of the constellation.) **5** ^{h} **20** ^{m}

E2.2 Use the Aim box to center Leo on the sky display. Estimate the right ascension of the name **Leo**. Write Leo's RA here: (**10** ^{h} **55** ^{m})

E2.3 Calculate the number of hours from Orion to Leo. Subtract Orion's RA (**E2.1**) from Leo's RA (**E2.2**).
$$(E2.2) - (E2.1) = \text{number of hours to wait}$$
$$(10^h \, 55^m) - (5^h \, 20^m) = (5^h \, 35^m)$$

E2.4 Now add that answer (**E2.3**) to 8:00 p.m. to find out when Leo will be high overhead.

4

(E2.3) + 8:00 p.m. = time when Leo is overhead
$$(5^h\ 35^m) + 8:00\ \text{p.m.} = (1:35\ \text{a.m.})$$

For questions Q21–Q24, follow the steps in the example just given. Suppose that Gemini is highest in the sky at 9:00 p.m., and you realize you wanted to look at something in the constellation Virgo. How late would you have to stay up in order to see the constellation Virgo at its highest point in the sky?

Q21 Click on the small arrow to the right of the Aim box, select **Constellations**, and click on **Gemini** to center the screen on this constellation. What is the approximate right ascension of the name Gemini?

Q22 Now center **Virgo** on the sky display in the same way. Write Virgo's RA here.

Q23 Calculate the number of hours from Gemini to Virgo. Subtract Gemini's RA (**Q21**) from Virgo's RA (**Q22**).
$$\textbf{(Q22)} - \textbf{(Q21)} = \text{number of hours to wait}$$
$$(\quad) - (\quad) = (\quad)$$

Q24 Now add that answer (**Q23**) to 9:00 p.m. to find out when Virgo will be high overhead.
$$\textbf{(Q23)} + 9:00\ \text{p.m.} = \text{time when Virgo is overhead}$$
$$(\quad) + 9:00\ \text{p.m.} = (\quad)$$

Q25 Use the Aim box to center on **Ursa Major**, the Big Dipper. Position your cursor over the end star in the Big Dipper's cup, Dubhe, and click on it to open its name tag. Click on the name tag to open Dubhe's information box, find the right ascension and declination, and write them here.

RA =   Dec =

Place the cursor over Dubhe, find Dubhe's altitude and azimuth from the status bar, and write them here.

Alt =   Azm =

Q26 Change the time in the Time settings panel by one hour, and press the <Enter> key on the computer keyboard. Now repeat the process to find Dubhe's right ascension declination and altitude azimuth coordinates one hour later.

Does the RA or Dec of the star change?

Does its Altitude or Azimuth change?

Q27 Look back at Q10. What is the declination of the zenith at your location?

Now look at the Location panel. What is your latitude?

Since the latitude of a place is its distance in degrees above or below the equator, and the declination of a star is its distance in degrees above or below the celestial equator, these two should be exactly the same.

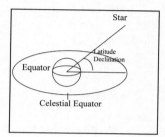

So we see that a point straight overhead (the zenith) will have the same declination as your latitude.

Q28 Remember that there are 90 degrees from the zenith down to the horizon. What will be the declination of the most southerly star (the lowest declination) that you might just be able to see at the horizon?

Start with the declination of the zenith, and subtract 90 degrees.
$$\text{Your lowest visible Dec} = \text{Dec of the zenith} - \mathbf{90}$$
$$= \mathbf{(Q27)} - \mathbf{(90)}$$
$$= (\quad) - \mathbf{(90)}$$
$$= (\quad)$$

Q29 Open the Find window, change the **Find:** box to **Stars, Proper names**, double-click on **Canopus**, and click **OK**. Now click on the star to open its name tag. Click on the name tag to open the information box. Find the declination of Canopus, and then close the box. Given your answer to Q27, would you ever be able to see the star Canopus from your location?

Section 4.5

Now we are going to find some of those picture-book objects. Select **Find** from the Information menu to open the Find window. Choose **Deep sky** in the **Find:** box and click on **Proper names**. Scroll through the list to locate each object in the data table on the following page. Double-click on the object name, and then click **OK** to rotate the view screen to center it. (You may have to increase the zoom to **10** to see some of these objects.) Click on the object to open its name tag, and then click on the name tag to open its information box. In the information box, find the right ascension and declination, and record them in the data table below. When you are finished recording the data, close the information box.

Now, repeat these steps until you have located all the objects in the data table.

Data Table

Deep-Sky Objects						
Object Name	RA	Dec		Object Name	RA	Dec
Andromeda Galaxy				Pleiades		
Dumbbell Nebula				Ring Nebula		
Lagoon Nebula				Sombrero Galaxy		
Omega Centauri				Triangulum Galaxy		
Orion Nebula				Whirlpool Galaxy		

If you would like see photographs of these objects, open the Photo Gallery from the Information menu. All of the galaxies will be found in **The extragalactic universe** section of the Photo Gallery under **Spiral galaxies**. The Dumbbell Nebula and the Ring Nebula can be found in **The Galaxy** section under **Planetary nebula**. The Lagoon nebula and the Orion nebula can be found in **The Galaxy** section under **Interstellar matter**. Omega Centauri can be found in **The Galaxy** section under **Globular clusters**. Finally, the Pleiades can be found in **The Galaxy** section under **Open clusters**. Remember that you can enlarge any picture to full-screen size by clicking on its Magnify icon and can go back to the picture list by clicking on the Thumbnail icon. When you have finished looking at the pictures, close the Photo Gallery window.

Section 4.6

When the data table is completed, insert the Extras CD in your computer's CD-ROM drive. When the small Start screen opens, click **Start** to open the Extras title page. In the Extras title page, click on **The Story of the Universe** button. In The Story of the Universe contents list, scroll down and select the item titled **The Celestial Sphere**. A movie window will open, and a narrated animation will begin playing. It is an excellent review of today's exercise. During the animation, there are some long pauses built in so that you can take notes.

At the bottom of the movie window is a horizontal "position" bar with a darker position indicator in it.

If you wish to see any part of the animation again, move the cursor to the approximate position of the part you want to see, and click. (Click at the left end to begin all over, in the center to begin near the middle, etc.)

When you have finished watching the narrated animation, click on the **Quit** button and proceed to the Discussion and Conclusion sections.

Discussion

Today you have learned how to describe an absolute address for a star or other celestial object in terms of right ascension and declination. Right ascension and declination are frames of reference fixed to the celestial sphere of stars and objects. As a result, anyone anywhere can use the right ascension and declination address of an object to find it.

Two things affect the right ascension and declination of stars: the Earth's *precession* and the *proper motion* of the stars. Both of these are extremely small, however, and are only factors when considered over many years, so we usually do not worry about them. We will look at these factors in a later exercise.

The other powerful aspect of right ascension and declination addresses is their ability to predict the positions of stars, constellations, and other objects at a later time or at a different location. The units of right ascension are based on the Earth's rotation. If you know the right ascension of a particular star, you can easily calculate when it will come up or go down if you know the right ascension of any object that is coming up or going down right now.

Similarly, declination is related to the Earth's latitude lines. If your observing site is 30 degrees north of the equator, then the stars that pass straight overhead at your site will also be at a declination of 30 degrees! And if you live in the Northern Hemisphere, the North Star, Polaris, will always be approximately the same number of degrees above the horizon as your latitude. Knowing your latitude, you can also predict whether or not you will be able to see a certain star just by knowing its declination.

Conclusion

Write a short summary describing the benefits of using right ascension-declination addresses for stars, constellations, and other objects. Include examples of how this address system can help your observing. Finally, give one example of a situation where you would *not* use RA-Dec but would use Alt-Azm instead.

4

Notes

Timing Is Everything

Introduction and Purpose

In the last several exercises, you learned how to locate things in the sky using *RedShift*. Along the way, you saw that finding stars and constellations depends as much on *when* you are looking as on where you are looking from. In today's exercise, we are going to look at the element of time. Take a moment and try to write a definition of the word "time."

Time is not an easy thing to define, but it is fairly easy to measure. One of the most obvious and easily measured passages of time is the 24-hour period we call a day. The Sun comes up—the Sun goes down—the Sun comes back up again! In today's exercise, you will find that the stars do the same thing.

Today's exercise has several purposes: to discover how the passage of time affects what we see in the sky, to discover how our location on the Earth relates to our measurement of time, and to identify which sky changes mark which time periods. You will also learn what determines time zones and how "local" time is different from "sun" time.

Procedure

Section 5.1

Start *RedShift College Edition*, bring the *RedShift 3* screen to the front, and open your personal settings file. Before you start today's exercise, you will need to turn off *RedShift's* Daylight Savings Time option. Select **Preferences** from the File menu. In the Preferences window, click on the **Follow system status** box to remove that check mark. Now click on the **Daylight savings time** box to remove that check mark, and click **OK** to close the Preferences window. Now, click on the date in the Time settings panel, and change it to today's date. Finally, change the Zoom factor to **0.7**, and press the keyboard <Enter> key to shrink the sky a little.

Section 5.2

Click on the Filters button, and select **Guides** to open the Guides filters window. In the **Local (Alt/Azm)** section of the window, click on the small button to the right of the **Horizon** box, and select **Line** to change the horizon to a single line. Then click on the **Zenith** box to turn that feature on.

In the **Celestial (RA/Dec)** section, click on the box to activate that section.

Make sure that the **Poles** box, the **Equator** box, and the **Numbering** box all have checks. If the **Grid** box is also checked, click on it to remove the check mark.

Now click on the **Ecliptic** box to activate that section. In the Ecliptic section, click on **Equator** and **Ecliptic plane** to turn those features **On**. If the Grid feature is checked, click on it to turn it **Off**. Finally, click **OK** to record your changes.

If the blue **NCP** label is not visible on the sky display, click on the <Up> or <Down> arrows on the keyboard until the label is just visible near the top of the screen. Now turn your attention to the Control Time panel.

Let's take just a moment to review the Control Time panel. The main component of this panel is a set of VCR buttons ◄ ◄◄ ■ ►► ► to start time moving forward or backward or to stop it. Underneath these buttons is a selection box titled **Step**. In our first exercise, we saw that this box determines how much time passes between each new screen display. The **Step** *number* can be changed by clicking on the number and typing in a new value. The **Step** *units* can be changed by clicking on the small arrow to the right of the box and selecting an appropriate option from the **Units** item of the drop-down menu.

With the time Step set at **1** hour, click repeatedly on the Forward Step button ►► and watch the screen change as the clock moves forward one hour each time. Now answer the following questions.

Q1 One star in the sky hardly moves at all. Which one is it? (Click on it to find its name if you don't already know it.)

Q2 All the other stars seem to move around the point labeled **NCP**. In which direction do they move, clockwise or counterclockwise?

Q3 Of course, the stars aren't really moving in a circle—this apparent motion is caused by the Earth's rotation. So in which direction is the Earth rotating, from east to west or from west to east?

Q4 Watch the end of the cup of the Little Dipper (Ursa Minor) as it moves. What practical use might these stars serve if you didn't have a watch?

Q5 Now click on the Forward Step button repeatedly, and count the number of hours it takes for the Little Dipper's cup to make one complete rotation

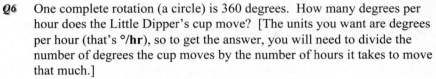

around the NCP.

Q6 One complete rotation (a circle) is 360 degrees. How many degrees per hour does the Little Dipper's cup move? [The units you want are degrees per hour (that's °/**hr**), so to get the answer, you will need to divide the number of degrees the cup moves by the number of hours it takes to move that much.]

$$(360°)/(\textbf{Q5}) = \text{number of degrees/hour}$$
$$(360°)/(\quad) = (\quad) °/\textbf{hr}$$

Q7 A clock face has 12 hours on it, so the hour hand on a clock moves all the way around in only 12 hours. How many degrees per hour does the hour hand on a clock move? (Solve the problem as in Q6.)

Q8 If the Little Dipper's cup was in a 7 o'clock position (pointing in the direction of the hour hand on a clock at 7 o'clock) just as it got dark, at what position would it be four hours later? (Remember that the Little Dipper rotates *counter*clockwise!)

(If you can't figure this out using your answers to Q6 and Q7, try it out with *RedShift*. Click the Forward Step button in the Control Time panel until the Little Dipper is in a 7 o'clock position. Then click on the Forward Step button four more times.)

Q9 Suppose you are out camping. Just as you are going to bed at 10 p.m., you notice that the Little Dipper's cup is pointing in a 3 o'clock position. You wake up some time in the middle of the night and see that the cup is now at a 12 o'clock position. What time did you wake up?
(As in Q8, you can use *RedShift* to verify your answer.)

Section 5.3

This motion of the stars is called daily motion because it repeats each day. Now change the direction of view to west by clicking on the <Left> arrow on the keyboard until the green **W** shows up on the horizon near the bottom center of the screen. Reset the date to today's date and the time to **8:30 p.m.**, and press the keyboard <Enter> key. Now answer the following questions.

Q10 Which constellation is nearest the upper left corner of the sky display?

Q11 Which constellation(s) is (are) just at the western horizon?

5

Q12 Now click on the Forward Step button. What happened to the constellation that was just at the western horizon (Q11)?

Q13 What happened to the constellation that was at the top left corner of the sky display (Q10)?

Q14 Now which constellation(s) is (are) just above the western horizon?

Q15 Click on the Forward Step button repeatedly, and count the number of screen updates until that same constellation is again just above the western horizon. Write the number here.

Q16 There is a red line and a blue line running through the horizon. Which one holds its position on the horizon as the hours pass?

Q17 One of these two lines is the *celestial equator* (a line running directly above the Earth's equator), and the other is the *ecliptic* (the path the Sun travels through the sky). Given your answer to Q16 and their positions on the horizon, which color line do you think is the *celestial equator*?

Section 5.4

As you probably guessed, your answer to Q15 was the same as your answer to Q5. The stars at the western horizon took the same amount of time to come back to their original positions as those near the celestial pole. However, instead of moving around in a circle, they were moving almost perpendicular to the horizon. They were "coming up" and "going down" just like the Sun.

Now let's see what the sky is like at a different time and place. Click on the **Local** button in the Time Settings panel and change it to **Universal**. Now change the time to read **00:00**, change the date to **01/01/2000**—midnight, January 1, 2000—and press the keyboard <Enter> key.

From the Controls menu, select **Choose Location**. Click on the button titled **Find sites** to open a window containing a list of cities. Look at the data table on the next page. Beginning with **Addis Ababa**, find each city listed below in *RedShift's* sites list, and select it. When the city is selected, click on the **OK** button to take you back to the previous window. Now click **OK** in this window to change your viewing location to that city.

Fill in the city's line in the data table. You will find the longitude and latitude in the Location panel below the city name. Round off the minutes and seconds to the nearest degree. (Remember, there are 60 minutes per degree.) If these values are labeled **E** (east), or **N** (north), record them as pluses; if they are labeled **W** (west) or **S** (south), record them as minuses.

Fill in the third column by dividing the city's longitude by 15° (*round it off to one decimal place*). To find the Local Time, click on the **Universal** button in the Time settings panel, and change it back to **Local**. The **Time Zone** of the city can be found just to the left of the Time box in the Time settings panel. If the Local Time is before midnight, December 31, 1999, then the Time Zone will be a negative number. If the local time is after midnight, January 1, 2000, then the Time Zone will be a positive number.

Finally, to find the time of Noon (transit), click on the small arrow to the right of the aim box, and select **Sun** from the drop-down menu to center the screen on the sun. Click on the Sun to open its name tag, and then click on the name tag to open its information box. In the information box, click on the **Report** button. In the Report window, you will find the times of rising, setting, and transit (or Noon) for the Sun. Write the transit time for this date in the last column, close the Report window, and close the information box. The first line is done as an example. Begin with the second line, and repeat these steps until the data for all the cities have been filled in.

Data Table

Cities of the World at Midnight, UT, 1/1/2000								
City	Lon.	Lat.	Lon. / 15°	Local Time	UT	Time Zone	Date	Noon (transit)*
Addis Ababa	+39°	+ 09°	+2.6	3 a.m.	0	+3	1/1/2000	12:28 p.m.
Bombay					0			
Calgary					0			
Denver					0			
Fairbanks					0			
Krasnoyarsk					0			
Perth					0			
Reykjavik					0			
Rio de Janeiro					0			

Because of the varying distance between the Earth and the Sun, our clocks are fast compared to the Sun in some parts of the year and slow in others. True noon on January 1 is around 12:03.

Q18 Look at the data table. What is the maximum variation in the time of noon (transit) between cities?

5

Q19   Which city has a noon time closest to 12:03? (Consider true noon to be 12:03.)

Q20   Is the time of noon in any city off by more than 30 minutes from true noon? (True noon is 12:03.)

Q21   Look at the data table. How does the **Time Zone** column compare with the **Long./15°** column?

Q22   In question Q6, you found that the apparent motion of the sky (caused by the Earth's rotation) is 15° per hour. If the Sun and stars move at 15°/hr, then how many hours would it take for the Sun to go from straight overhead at Addis Ababa (38 degrees east of Greenwich, England) to straight overhead at Greenwich?

Q23   On the basis of your answers to the previous two questions, how wide in degrees of longitude is one time zone?

Section 5.5

As you can see from the Noon column in the data table, our time zones are set up so that the time in each location is fairly close to Sun time. Noon ought to be when the Sun is highest in the sky no matter where on the Earth you are. This is the reason for time zones. Of course, each community could establish its own real time based solely on the Sun, but then you would never know what time it was when you traveled. So we compromise and establish time zones every fifteen degrees of longitude.

We center these time zones on lines called *meridians* that run from the North Pole to the South Pole. These *meridians* are placed every fifteen degrees of longitude, beginning with the *Prime Meridian* running through Greenwich, England. This way any city exactly on a meridian (whose longitude was an exact multiple of 15) would find its time approximately synchronized with the Sun. (The Earth's varying distance from the Sun through the year makes even this relationship only an approximation.)

Even if you are not on a meridian, however, you can determine what your approximate Sun time is. Re-open your personal settings file as at the beginning of this exercise. Set the **Local** date and time to **12:00 p.m.** (noon), **06/14/2000** and press the keyboard <Enter> key. Now answer questions Q24–Q28 to learn how to calculate the approximate Sun time for your particular location.

Q24 Divide your longitude by 15°. Write down *only the decimal* part of the answer.

Q25 Fifteen degrees represents 60 minutes of time, so multiply your answer to Q24 by 60, and *round it to the nearest whole number* to get the number of minutes your clock is off from Sun time.

Q26 If your answer to Q25 was 30 minutes or greater, you are more than halfway to the next meridian and therefore are in the next time zone, so subtract your answer to Q25 from 60, and write that number here.

If your answer to Q25 was less than 30 minutes, leave it as is.

If your longitude is **West**, answer Question 27.

Q27 If your answer to Q25 was 30 minutes or greater, add Q26 to your clock time (12:00) to get the approximate Sun time, and write your answer here.

If your answer to Q25 was less than 30 minutes, subtract your answer to Q26 from your clock time (12:00) to get the approximate Sun time, and write your answer here.

If your longitude is **East** answer Question 28.

Q28 If your answer to Q25 was 30 minutes or more, subtract your answer to Q26 from your clock time (12:00) to get the approximate Sun time and write the answer here.

B) If your answer to Q25 was less than 30 minutes, add your answer to Q26 to your clock time (12:00) to get the approximate Sun time and write your answer here.

Now here's what all this means. If your Sun time (Q27 or Q28) turned out to be several minutes before noon, then the Sun has not yet reached its highest point when your clock reads 12:00 and will not go overhead until that many minutes afterward. Conversely, if your Sun time was several minutes after noon when your clock read 12:00, then the Sun had already passed overhead that many minutes before. Click on the small arrow to the right of the Aim box, and select **Sun** to center the Sun on the screen. Open the Sun's information box and Report window to find the time of transit—the actual time the Sun was overhead. It should be off from 12:00 by an amount equal to the answer to Q26.

Section 5.6

The last thing we wish to do today is to figure out how to calculate the time in one place if we know the time in another. Notice in the data table that you can obtain the correct time zone simply by rounding off the result in the Longitude/15° column *to the nearest whole number.* After thus calculating the time zone for each place, you can find the difference in time between them by finding the number of time zones between the two and then adding or subtracting that many hours from the known time. To see how this works, study Example 1 below. Then answer the following questions.

Example 1: If it is 2 p.m. on June 20, in Boise (Longitude 116.2° West), what is the date and time in Berlin (Longitude 13.42° East)?

    E1   First we must find the time zones for each place: Divide each longitude by 15°, and round it off to the nearest whole number. *(Remember that East longitude is + and West longitude is −.)*

        A)   Boise: $-116.2°/15° = -7.75$ rounded off to -8

        B)   Berlin: $+13.42°/15° = 0.90$ rounded off to $+1$

    E2   Next, we find the difference between (subtract) the two time zones, *beginning with the most eastern time zone.*

$$\text{Berlin} - \text{Boise} = +1 - (-8)$$
$$\text{Time zone difference} = 1 + 8$$
$$= 9$$

    E3   The city with the greatest time zone number (the farthest east) is the one with the latest time. (Just remember "*greatest is latest.*") In this case, Berlin's +1 is greater (later) than Boise's −8, so Berlin's time is 9 hours *later* than Boise's.

        Since we know it is *later* in Berlin than in Boise, we *add* the nine hours to Boise's time to get Berlin's.

$$\text{Boise time} + 9 \text{ hours} = \text{Berlin's time}$$
$$2 \text{ p.m.} + 9 \text{ hours} = 11 \text{ p.m., Berlin time (still June 20, but barely!)}$$

Follow the above example to answer the following questions.

    Q29   If it is 3 p.m., March 22, in Calcutta (Longitude 88.3 degrees East), what is the date and time in Kiev (Longitude 30.4 degrees East)?

Q30 If it is 7 a.m. on June 1, in Tokyo (Longitude 139.75 degrees East), what is the date and time in Fairbanks (Longitude 147.8 degrees West)?

After you have found the answers, you may check your work. As before, use *RedShift's* sites list to set the sky to the first location. Set the Local time and date to the values given in the problem, and press the keyboard <Enter> key. Then go back to the Sites list, select the other location, and see what time it is there.

Section 5.7

Now you need to reset your preferences. Select **Preferences** from the File menu again. Click on the **Daylight savings time** box to turn that function back on, and then click on the **Follow system status** box to turn it back on. Finally, click **OK** to close the Preferences window.

Now insert the Extras CD in your computer's CD-ROM drive. When the small Start screen opens, click **Start** to open the Extras title page. In the Extras title page, click on **The Story of the Universe** button, and in the contents list, scroll down and select the item titled **Time and Seasons**. A narrated animation describing time keeping will begin playing. This animation is an excellent summary of this exercise. Watch the narrated animation up to the part about the Julian Date.

As in the last exercise, you can replay any part of the animation by moving the cursor over the horizontal position bar at the bottom of the window to the approximate position of the part you want to see again and clicking.

(Click at the left end to begin all over, in the center to begin near the middle, etc.) Once you have watched the animation to the Julian Date section, you can click on the **Quit** button to close the movie window.

Discussion

Today, you have learned how the sky seems to move as the Earth turns. This motion repeats every 24 hours and is called daily motion. Since the sky rotates through a complete circle of 360 degrees each 24 hours, we found that it moves 15 degrees per hour. This means that if a certain constellation is overhead at 8 p.m., a constellation 15 degrees further east will be straight overhead at 9.

It also means that if a certain constellation (or celestial object like the Sun or Moon) is straight overhead at one place on the Earth at 8 p.m., it will be straight overhead at a place 15 degrees further to the west at 9 p.m.

5

Thus, we found that measurements of longitude on the Earth correspond to measurements of time in the sky. In the opposite sense, an event that occurs at a particular *time* in the sky will be seen only at a particular *place* on the Earth. This is why an eclipse, for instance, can be seen in only one particular area of the Earth while everyone else misses it.

Finally, we found that time zones allow us to keep our sense of time fairly closely aligned with the sky (or at least with the Sun). We also learned how to determine the time zone of a particular place just by knowing its longitude, how to find real Sun time by knowing our longitude, and how to use time zones and longitude to determine what time it is in other places on the Earth. In the next exercise, we will learn even more about time-keeping and the sky. We will take a look at a couple of other celestial time periods, the month and the year.

Conclusion

In a short paragraph, describe the effect of the rotation of the Earth on what we see in the sky. Describe how you might use the Little Dipper as a clock at night if you didn't have a watch. Tell how fast the sky seems to move and how to use that information to predict the position of various celestial objects at different times or at different places on the Earth. Also, describe the use of time zones on the Earth and how to tell what time it would be somewhere else just by knowing your longitude and the longitude of the other location. (Continue your conclusion on a separate sheet of paper if necessary.)

Time Passages

Introduction and Purpose

Since the beginning of "time," humans have measured (and to some extent defined) time in terms of celestial events. In the last exercise, we examined the most basic time passage, the **rotation** of the Earth. We saw the effects of this 24-hour period of rotation on the sky as well as on our clocks. But there are other time passages that are just as important to us. One year, for instance, is the time required for the Earth to go around the Sun once.

One way in which the ancients measured this period of time was by observing the **heliacal rising** of a bright star. They would watch each morning just before sunrise to see if they could spot some bright star. One morning, the star would just be visible before the sun's light washed it out. This was the date of the star's **heliacal rising**.

Each morning thereafter, the star would be higher and higher in the sky at the same time and thus easier and easier to see. After many days, the star would work its way all the way across the sky and be down by sunrise. When 365 days had passed, the star would again be spotted just before dawn in the eastern sky (another **heliacal rising**), and the ancients knew that one year had passed.

In today's exercise, we will examine the effects of the Earth's **revolution** around the Sun—the "annual motion." We will see how this motion affects the sky through the months and how the ancients set up a crude "calendar" in the sky called the **Zodiac**.

Procedure

Section 6.1

Start *RedShift College Edition*, and then bring the *RedShift 3* screen to the front. When the *RedShift 3* screen opens, click on the File menu, and open your personal settings file. The only thing we will need to change is the date. Click on the **Date** in the Time settings panel and change it to today's date. Change the Zoom factor to **0.7**, and press the keyboard <Enter> key to shrink the sky a little.

6

Section 6.2

To review the "daily motion" we examined in the last exercise, leave the Step set at 1 hour, and click the Forward Step button repeatedly. Observe the motion of the constellations, and watch the sky move through one whole rotation (24 steps). Now, let's change the time step. In the Control Time panel, click on the small arrow to the right of the Step box, select **Units**, and click on **Day**. Now click on the number, type in **30**, and press the keyboard <Enter> key.

Imagine that you went out to look at the stars only one night out of every 30 at the same time of the evening. How would the star positions have changed since your last observation? The computer is now set to re-calculate the star positions every 30 days. Now answer questions Q1–Q5.

Q1 Click the Forward Step button repeatedly. How do the star positions every 30 days compare with the star positions every hour (the "daily motion")?

Q2 Is this new star motion in the same direction as the daily motion was, or in the opposite direction?

Q3 Watch the position of the Little Dipper. How many 30-day intervals are required before the Little Dipper returns to the same position at the same time of night?

Q4 About how many days (total) does it take for the Little Dipper to come back to the same position in the sky at the same time of night? (Multiply your answer to Q3 by 30 days to get the total.)

  (Q3) × 30 days = total days to return to the same position
  (   ) × 30 days = (   ) days

Q5 What standard time period is your answer to Q4 similar to?

Section 6.3

Now change the sky settings as shown in the table below, and press the keyboard <Enter> key.

Local time	8:30 p.m.
Date	01/07/2000
Step	30 days
Zoom	0.4

Click on the small arrow to the right of the Aim box, and select **Sun**. Then click on the tiny **Lock Aim** button (as shown) to lock the Sun in the middle of the screen. The Sun should now be a small whitish disk centered on the screen.

Finally, from the Control menu, select **Base Plane**, and click on **Ecliptic plane** to change the orientation of the screen. The sky will rotate so that the ecliptic (the path of the Sun) is horizontal with the screen. The ecliptic will be a red line going more or less horizontally across the center of the celestial sphere. Now answer the following questions.

Q6   Look at your display and write down the constellation the Sun is in to begin with.

Q7   With the Step still set at **30 days**, click the Forward Step button repeatedly. Count the number of screen changes until the Sun is back in the same constellation.

Q8   How many days (total) does it take for the Sun to come back to the same constellation?
(Multiply your answer to Q7 by 30 days to get the total.)
       (Q7) × 30 days = days to return to the same constellation
       (   ) × 30 days = (   ) days

Q9   Your answer to Q8 should look familiar! The length of time required for the Sun to move through the sky and return to its beginning point is, of course, one year. Did the Sun return to exactly the same spot in the constellation as it started from?

Section 6.4

In the previous simulation, we could have chosen a Step other than 30 days that would have given us a closer answer to the correct length of a year. As you noted, the Sun did not quite return to the same place it started from in 360 days, but it was close. We chose 30 days for a reason, however. Watch the following simulation, and see if you can deduce what it is.

First, turn off everything in the Display menu except **Sun & Planets**, **Moons**, and **Stars**. Select **Base Plane** from the Control menu, and click on **Eq. Earth** to rotate the screen back parallel to Earth's equator.

Now, we need to select which planets are visible. Click on the Filters button and select **Sun & Planets**. In the Sun & Planets filters window,

6

click on the **Hide** buttons (the very first column of buttons) for all the objects *except* the Sun. Then click on the Moons tab at the left of the window. First, make sure the Earth's Moon is set to show an image. Then click on the tab titled **Paths** to leave a trail behind as the Moon moves through the sky. These settings will turn off everything except the Sun and Moon and cause the Moon to leave a trail showing its path. Click **OK** at the bottom of the Filters window to record these changes.

Click on the **Local** button in the Time settings panel and change it to **Universal**. Now change the sky settings as shown in the table below, and press the keyboard <Enter> key.

Universal time	6:00
Date	01/09/1997
Step	1 day
Zoom	0.26

At this point, you should see a whitish disc (the Sun) in the middle of the screen and a slightly smaller disc (the Moon) right above it. When the Moon is lined up with the Sun like this, we call it a New Moon. Place a piece of transparent tape on the monitor to mark the original position of the Moon.

Now click on the **Lock Aim** button again to *unlock* the Sun. Click the Forward Step button, and count each click as the Moon goes across the screen to the left and then reenters the screen from the right. The Sun will also be moving slowly to the left. Continue clicking (and counting) until the Moon passes the newest position of the Sun (not the original position). When you have finished the simulation, answer the following questions.

*If you lost count (or missed a setting), reset the date and time (remembering to press the <Enter> key). Then click on the **Filters** button, and select **Sun & Planets**. When the Filters window opens, click **OK** to close it again. This action will erase the Moon's "footprints." Finally, use the Aim box to re-center the screen on the **Sun**, and repeat the simulation.*

Q10 Did the Moon come back exactly to the same alignment with the Sun as it started from?

Q11 What was the number of the last click just before the Moon passed the Sun? (Subtract one from your total number of clicks. This is the number of complete days between New Moons.)

Q12 Approximately what fraction of a day would need to be added to Q11 to bring the Moon back to its original alignment directly above the Sun? Add this fraction to Q11 to find the total time between New Moons. (The final answer should be something like 20 1/2 days or 26 1/3 days.)

Now you need to clear the Moon's footprints so you can repeat the simulation. To erase any paths currently on the screen, click on the **Filters** button, and select **Sun & Planets**. When the Filters window opens, click **OK** to close it again. This action will erase any footprints currently on the screen.

Q13 Repeat the preceding simulation, this time counting clicks (days) only until the Moon returns to *its* original position (marked by the piece of tape). How many days did it take for the Moon to return to *its* beginning point? (Again, include any fraction of a day that you can estimate.)

Now you know why we chose 30-day intervals to divide the year. Thirty days is approximately the length of a month, from New Moon to New Moon and divides equally into 360. So (of course), there are approximately 12 such "lunar" months in a year.

The length of the "lunar" month depends on how you define it. If you define it with respect to the Sun, from New Moon to New Moon, it is called a *synodic month* and has a length of approximately 29 1/2 days. This is the way most cultures have traditionally observed the month. If you define it with respect to the rest of the sky and the stars, it is called a *sidereal month* and has a length of approximately 27 1/3 days.

Section 6.5

We now turn our attention back to the "annual motion" of the sky. Reopen the Moons filters window from the Filters button, click on **Paths** once more to erase the Moon's path, and click **OK**. Click on the small button to the right of the Aim box, and select **Sun**. Now click on the **Lock Aim** button to again lock onto the Sun.

In the Display menu, turn the Constellations and Guides back on. Now change the sky settings as shown in the table below, and press the keyboard <Enter> key.

Local time	6:00 a.m.
Date	01/07/1999
Step	30 days
Zoom	0.7

In what constellation will the Sun be on that day? Write the constellation name in the January line of the 1999 AD column in the data table on the next page. Now, click on the Forward Step button. One month later, which constellation is the Sun in? Write the name of this constellation in the February line of the 1999 AD column. Continue like this until you have filled in the first column of the data table.

These 12 constellations through which the Sun passes are known as the "signs of the Zodiac." They were devised long ago, thousands of years BC, probably over a long span of time. Many people still believe that these "signs" and the meanings given them by the ancients control their destinies. The way in which the old meanings are applied today is to see what sign is in effect on the person's birthday. But is the Sun really in your sign on your birthday? Not any more!

You see there is one more motion of the sky as seen from the Earth that takes many thousands of years to become apparent. It is called *precession*. The angular momentum of the Earth's spin causes the North Pole to rotate very slowly around the sky, taking 26,000 years to complete one circle. There are several effects of *precession*. We will look at one of them today.

The constellation where the Sun is today on a given date is not the constellation where the Sun was when the signs of the Zodiac were originally used! To see the effect of *precession* on the signs of the Zodiac, set the date in the Time settings panel to **01/07/0004 BC**, and repeat the above procedure to find the constellation the Sun was in during each month of that year.

Finally, set the date to **01/07/2000 BC** (when many of these constellations were being invented) to see where the original Zodiac "astrology" came from. When you are finished, answer the following questions.

Data Table

Constellations Housing the Sun for Different Months			
Month	1999 AD Constellation	4 BC Constellation	2000 BC Constellation
January			
February			
March			
April			
May			
June			
July			
August			
September			
October			
November			
December			

Q14 What constellation was the Sun in on March 21, 1999, AD (the spring equinox)?

Q15 What constellation was the Sun in on March 21, 2000, BC?

Q16 What constellation was the Sun in on September 23, 1999, AD (the fall equinox)?

Q17 What constellation was the Sun in on September 23, 2000, BC?

Q18 According to astrologers, a person born in January, 2000, BC, was supposed to have certain personality and character traits because they were born when the Sun was in Aquarius. Does it make sense that a person born in January, 1999, AD (when the Sun is in Sagittarius) would also have the personality and character traits of Aquarius?

Q19 Given the changes in the constellations from 2000 BC to 4 BC to 2000 AD, in what constellation would you expect the Sun to be during the month of January, 4000, AD?

Q20 Archaeologists (like Indiana Jones) like to look for celestial alignments in ancient buildings and worship sites. It seems apparent that people did at

6

least occasionally try to align their buildings, temples, and ceremonial drawings with important rising or setting points of the Sun, bright planets, or notable constellation stars. Considering what you now know about precession, what problem arises in trying to use such alignments today? (Could Indiana Jones really have used that crystal-on-a-stick in the movie *Raiders of the Lost Ark*?)

Q21 Change the date to **12/14/1997 AD**. In what constellation is the Sun on this day?

Q22 Is this constellation (Q21) one of the 12 Signs of the Zodiac?

Section 6.6

If you click on the **Local/Universal** button in the Time settings panel, you will see a third option—**Julian Date**. To find out what the Julian date is (and review annual motion), insert the **Extras** CD in your computer's CD-ROM drive. When the small Start screen opens, click **Start** to open the Extras title page. In the Extras title page, click on **The Story of the Universe** button, and in the contents list, scroll down and select the item titled **Time and Seasons**. A movie window will open, and a narrated animation describing time keeping will begin playing. Click in the center of the horizontal position bar at the bottom of the movie window to jump to the section on the Julian date. Watch the rest of the narrated animation. As always, you can replay any section by clicking on the position bar near where you wish to begin.

Q23 In a few words, describe the Julian date system.

Discussion

The Earth's motion around the Sun has given rise to many important celestial observations. Because of the tilt of the Earth's axis, this annual motion results in the seasons. The need to know when the seasons were about to change prompted the ancients to very carefully record the four observable points in the Earth's motion around the Sun—the two equinoxes and the two solstices.

It has been proposed that the twelve constellations of the zodiac were originally only four—one for each of those important solar positions. Then, in a few thousand years (thanks to precession) the four principal solar positions moved out of these original constellations, so four more had to be invented to correspond to the new solar points. Eventually, after Earth's precession had moved the equinoxes and solstices into four new areas of the sky, a final four were put together to keep track of the Sun through the year, resulting in a final total of 12 zodiacal constellations.

Do a little test on the system of astrology. The next time you read a newspaper, check out the horoscope. Get a friend or two to listen to you read their horoscopes, but instead of reading the correct horoscope for each one's birthday, read the one for a month or two before or after their birthdays. How many items did they find that fit them?

Another interesting test is to cut out each month's horoscope leaving off the name of the sign. (Make a note somewhere so that you will know which horoscope goes with which sign.) Then mix them all up and have a friend pick the one he thinks fits him best, and see just how well astrology really works.

Conclusion

In a few paragraphs, describe how the Earth's annual motion around the Sun affects which constellations we see in the evening sky. Tell which constellations are associated with which seasons. Be sure and mention in which direction the constellations seem to move (eastward or westward) during the year. Also mention how precession affects the constellations associated with each season.

Optional Observing Project

Do the observation exercise "The Moon" in Appendix C.

Notes

Eclipses

Introduction and Purpose

While the stars are beautiful and awe-inspiring, humans have always been most impressed by the two great lights in the sky, the Sun and the Moon. In previous exercises, we observed the Sun's motion through the sky during the year, as well as the period of the Moon with respect to the Sun (the *synodic* period). In today's exercise, we will see how the combination of these motions creates the phases of the Moon. We will also examine the Moon's *sidereal* period (its period with respect to the stars) and watch its path through the sky during a month. The Moon also occasionally teams up with the Sun to produce two of the most feared celestial effects in history—solar and lunar eclipses. We will see how to use *RedShift* to predict these events.

Procedure

Section 7.1

Start *RedShift College Edition*, and then bring the *RedShift 3* screen to the front. When the *RedShift 3* screen opens, click on the File menu and open your personal settings file. In the Display menu, click off everything *except* **Sun & Planets**, **Moons**, and **Stars**. In the Control menu, select **Base Plane**, and then click on **Ecliptic plane** as shown. This setting will keep the screen parallel to the path of the sun through the sky. Now, click on the **Local** button in the Time settings panel, and change it to **Universal**. Finally, change the sky settings as shown in the table below.

Universal time	**00:00**
Date	**12/08/1999**
Step	**12 Hours**
Zoom	**0.26**

Click on the filters button, and select **Sun & Planets** to open the Sun & Planets filters window. Click on the large **Hide** button to hide the Sun and all the planets. Now, click on the **Moons** tab at the left of the window to open the Moons filters. Make sure the fourth button (**Image**) is clicked **On** for the Moon. Then click **OK**.

Now click on the small arrow to the right of the Aim box, select **Stars**, and then click on **Antares**. This will rotate the screen to center on the bright star Antares, which is very near the Moon on that evening. Click on the small **Lock Aim** button just above the Aim box to lock the screen on Antares. You should see a small white circle representing the Moon near the center of the screen. Place a piece of transparent tape on the monitor to mark the original position of the Moon. Then answer the following questions.

Q1   Click on the Forward Step button, and count clicks as you watch the Moon move across the sky. The Moon will move all the way off the left side of the screen and disappear. Continue to click and count until the Moon reappears on the right side of the screen. When it has returned to its starting point, you may stop. What was the exact number of clicks for one complete revolution of the Moon around the Earth? (Estimate the fraction of the last click needed to come back exactly to the starting point.) Write the number here.

Q2   According to the Time step box, each interval is 12 hours. This is one-half of a day, so divide your answer to Q1 by 2 to find the ***sidereal period*** of the Moon (the period of the Moon's motion with respect to the stars).
Period of the Moon   = **(Q1)/2**
= (   )/2
= (   ) days

Section 7.2

Now, let's take a closer look at the Moon as it goes through one month. From the Control menu, select **Base Plane**, and click on **Ecliptic plane** to rotate the view. Now let's unlock the screen by clicking on the **Lock Aim** button. Then click on the small arrow at the right of the Aim box, select **Planets/Moons**, choose **Earth's moon(s)**, and click on **Moon** to center the screen on the Moon. Now click on the **Lock Aim** button again to lock it there. Finally, change the sky settings as shown in the table below.

Universal time	**00:00**
Date	**12/08/1999**
Step	**12 Hours**
Zoom	**10**

Now you are going to see how the appearance of the Moon changes as it moves around the sky.

Q3 Look at the brightness level of the Moon. Would you be able to see the Moon if it were completely dark like this?

Q4 What phase of the Moon is this (when it cannot be seen)?

Now click on the Forward Play button in the Control Time panel, and watch as the days pass. See how the Moon's apparent shape and brightness change. Stop the simulation when the Moon is half illuminated, and record the date here. This phase of the Moon is called *First Quarter*.

Q5 How many days have passed since New Moon?

Click on the Play button again, and watch until the Moon is full. Stop the simulation again, and record the date here.

Q6 How many days have passed since First Quarter Moon?

Q7 How many days have passed since New Moon?

Section 7.3

Now open the Sun & Planets filters window, make sure the fourth button (**Image**) next to the Sun is checked, and click **OK**. This setting will add the Sun to the sky display. Change the sky settings as shown in the table below and then answer questions Q8–Q13.

Universal time	**00:00**
Date	**12/08/1999**
Step	**1 day**
Zoom	**0.5**

Q8 On December 8, 1999, the Sun and Moon are almost in the same spot in the sky. This is called a New Moon. You can't see the Moon on this day because it is up when the Sun is up and down when the Sun is down. Do they overlap, or is it a close miss?

Q9 Click the Forward Step button repeatedly (counting your clicks), and watch the Moon and the Sun move across the sky. This time, continue until the Moon has gone all the way off the screen, reappeared on the right, passed its starting point, and caught back up with the Sun (which has moved some distance to the left during this period). Then stop the simulation.

You have now arrived at a second New Moon. How many intervals (days) passed during one complete cycle of the phases of the Moon—from one New Moon to the next? (Estimate the fraction of the last interval (day) needed to come back exactly to the second New Moon.) As we mentioned in the last exercise, this number is called the *synodic period* of the Moon. Write the number here.

Q10 Click on the Filters button and select **Guides** to open the Guides filters window. In the Guides filters window, change the Horizon setting to **Line**. Click the **Ecliptic** box to activate that section, and then make sure that **Ecliptic plane** is selected and the **Grid** is turned **Off** as shown. Now click **OK**, and from the Display menu, turn on the **Guides** option. The red line that appears on your screen is the path of the Sun, called the ecliptic. Once again, reset the sky display to the settings shown below.

Universal time	00:00
Date	12/08/1999
Step	1 day
Zoom	0.5

Click on the small **Lock Aim** button to release the lock on Antares. Now click on the small arrow to the right of the Aim box and select **Planets/Moons.** Then select **Earth's Moon(s)**, and click on **Moon** to center the screen on the Moon once more. Click on the **Lock Aim** button again to lock the screen on the Moon, and repeat the simulation.

Now you should notice that the path of the Moon does not exactly coincide with the path of the Sun (the ecliptic). Watch for the point where the ecliptic is farthest from the Moon, and stop the simulation at that point.

Now open the Tools menu and select the **Ruler** tool. Once the Ruler tool is open, move the mouse until the pointer is on the ecliptic (either directly above or directly below the Moon), and click the left mouse button. Now move the mouse until the pointer is right in the center of the Moon and click the left mouse button again. You can now read the angle between the ecliptic and the center of the Moon in the data box titled **Result angle.** Write that maximum value here.

Q11 Click on the Tools menu, and again click on **Ruler** to turn that function back off. Continue the simulation until the Moon passes the Sun. Now click the Reverse Step button to move the Moon back past the Sun. When the Moon passed the Sun again, did it look like the Sun and Moon overlapped, or was it another close miss?

Q12 Let's verify your answer to the previous question. Change the Zoom to **10** to see the Moon and the Sun considerably enlarged. Change the Step in the Time settings panel to **1 hour**, and single-step forward or backward to get the Sun and Moon as close to overlapping as possible. Does the Moon go above the Sun, in front of the Sun, or below the Sun?

Q13 Now change the sky settings as shown in the table below.

Step	**1 day**
Zoom	**0.26**

Click on the **Lock Aim** button again to release the lock on the Moon. Using the Aim box, select **Stars**, click on **Antares**, and lock it back in the center of the screen. Now click on the Forward Step button repeatedly and watch till the Moon goes all the way across the screen and once again catches up with the Sun. Stop the simulation at this point again.

As before, change the Zoom to **10**, and use the Aim box and **Lock Aim** functions to lock the Moon in the center of the screen. Change the Step to **1 hour**, and single-step back and forth to locate the exact point at which the New Moon occurs (when the Moon and Sun are right together). This time, does the Moon go above the Sun, in front of the Sun, or below it?

Section 7.4

You saw that the Moon just grazed above the Sun-Earth line on January 6 and just grazed below the Sun-Earth line on February 5. What do you think the Moon's position will be halfway between those two dates? Let's check it out.

First, click on the **Lock Aim** button to release the lock on the Moon. Now change the sky settings as shown in the table below.

Universal time	**20:00**
Date	**01/19/2000**

Step	**6 Hours**
Zoom	**0.26**

Click on the Filters button, and select **Sun & Planets** to open the Sun & Planets filters window. Click the fourth button (**Image**) next to Earth. Then click **OK**. This will turn on the Earth image and show the surfaces of the Sun, Moon, and Earth as they would look under actual sky circumstances. Now, we are going to use a couple of new *RedShift* functions. First, find the Location panel. Notice the large button just below the panel name initially titled **Observe from surface**. Click on this button, and select **Space: track body** from the menu that drops down.

When the screen has reset, you should see an image of the Earth in the center with the Moon on one side and the Sun on the other. Now let's rotate the view. If the Lon (longitude) and Lat (latitude) boxes are not visible in the Location panel, click on the small icon at the upper right corner of the Location panel 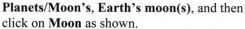 to expand it. Now click in the Lat (latitude) and Lon (longitude) boxes, change the values as shown below, and press the <Enter> key on the computer keyboard to rotate the view.

Lon	**00°, 00′, 00″**
Lat	**000°, 00′, 00″**

Now you should just be able to see the Moon a little to the right of the Earth. Answer the following questions.

Q14 Click repeatedly on the Forward Step button. What happens to the Moon?

Q15 As you look at the Earth in this picture, the Sun is over your shoulder and directly behind you, shining straight on the Earth. While the Moon is behind the Earth, how much sunlight will fall on the Moon?

Section 7.5

Now let's see what is happening on the Moon. Click on the small arrow to the right of the top Location box, select **Planets/Moon's**, **Earth's moon(s)**, and then click on **Moon** as shown.

This setting will move the focus from the Earth to the Moon. Now change the sky settings as shown in the table below.

Universal time	**2:08**
Date	**01/21/2000**

Step	5 Minutes
Zoom	1

You should see a small, dark shadow just touching the left edge of the Moon's surface. This is the point where the outer edge of the Earth's shadow first touches the Moon and is called the ***first contact***. The outer part of the Earth's shadow is called the ***penumbra***. Now answer the following questions.

Q16 Click on the Forward Step button seven times and watch the penumbra shadow move across the face of the Moon. In which direction is the shadow moving?

Q17 Is the shadow across the face of the Moon straight or curved?

Q18 What could you deduce from your observation in the previous question about the shape of the object casting the shadow? (Aristotle understood this fact 300 years before Christ was born.)

Q19 Continue clicking the Forward Step button until the shadow covers the Moon. Continue to click the Forward Step button, watching the left edge of the Moon for an even darker shadow. This part of the Earth's shadow is called the ***umbra***. This is the ***second contact***. Note the Universal time when this occurs. Write the time here.

Q20 Now continue the simulation and watch until the darker shadow has also covered the Moon completely. Again, write the time here.

Click on the small arrow at the right of the top Location box, select **Planets/Moons**, and click on **Earth** to move back to the other side of the Earth. As before, you're looking at the sunlit side of the Earth, and the Moon is hidden behind the Earth in its shadow.

Q21 What countries or continents are facing the Moon when the Moon is behind the Earth?

Q22 Will anyone in Australia be able to see this eclipse?

Q23 Once more, click on the small arrow at the right of the top Location box, choose **Planets/Moons**, select **Earth's moon(s)**, and click on **Moon** to change the focus back to the Moon. Now continue to click the Forward Step button until the Moon begins to emerge into the lighter shadow again. Note the time and write it here.

Q24 Continue the simulation until the darkest shadow is completely gone. Note the time and write it here.

Q25 Continue clicking the Forward Step button until the Moon is completely illuminated again. Write down the time.

Q26 How many different *times of contact* were there in the eclipse (from the very beginning of the eclipse to the very end)?

Section 7.6

Now, from the Events menu, select **Eclipses**. The Eclipse window is a very powerful tool that will locate and describe all eclipses during any period of time. Since we are currently looking at lunar eclipses, click on the small arrow to the right of the Search type box, and select **Lunar eclipse**. So as not to create too large a search, set the beginning date to **1997** and the ending date to **2001**. Then click **Search.** The computer will go to work and list each lunar eclipse between the selected dates.

Locate the eclipse of **January 21, 2000**, in the list, and click on it to list the **Circumstances of the eclipse** in the lower right-hand part of the window. Then answer the following questions.

Q27 Look at the times listed in the Circumstances box, and then look at the times you recorded in the previous questions. What is the time of second contact (Q19) called in the Circumstances section?

Q28 Which contact is called **Moon enters totality**?

Q29 Now click on the **OK** button. When the screen has finished resetting, click once on the Forward Step button. Now, describe what point in the eclipse you are viewing.

Q30 From the Information menu, choose **Photo Gallery.** In the Contents frame, click on **Solar System**, and select **Moon.** (If the Contents frame is not open, click the **Contents** button in the navigation bar to open it.) In the Moon photo window, scroll all the way to the bottom. Find the photo titled **[3025] Total lunar eclipse, 18/19 November 1975**, and enlarge this photograph of a total eclipse of the Moon. Here, you can see all three parts of the eclipse. The umbra is on the left, the penumbra is in the middle, and the noneclipsed moon is on the right edge. Notice the beautiful (and weird) color. Click on the Thumbnail button, read about the eclipse, and describe what causes this color.

Close the Photo Gallery window by clicking on the small close button at the upper right corner of the Photo Gallery. (Be careful not to close the entire *RedShift* window.) *(Macintosh users will click the small square in the upper left corner of the window to close it.)*

Section 7.7

Now, let's turn our attention to Solar eclipses. Again, choose **Eclipses** from the Events menu, but this time, select **Solar eclipse** in the Search type section. In the Search interval box, enter dates from **2010** to **2020**. Make sure the **Search for current location** box does *not* have a check mark in it, and click **Search**. When the search is finished, select the eclipse for **August 21, 2017**, and click **OK**. This will set the sky to the beginning of the next major total solar eclipse that will be visible from most of North America, near the point of longest totality. The location is set for just northwest of Nashville, Tennessee. Now answer the following questions.

Q31 In this view, the eclipse is well under way. To go back to the beginning of the eclipse, click the Reverse Step button until the Moon just contacts the Sun's edge. Write down the Universal time.

Q32 Now, let's go forward in time. Click on the Forward Step button until the last little bit of the Sun almost disappears. Now change the time Step in the Control Time panel to **1 minute**. Click on the Single Step buttons forward or backward to locate the exact beginning of totality, when the Sun is first completely blacked out, and write that Universal time here.

Q33 Click on the Forward Step button repeatedly until the Sun's edge just pops out on the other side. Write the Universal time of this event here.

Q34 How long did the totality last? (Subtract Q32 from Q33.)

Section 7.8

Now, let's see what this eclipse looked like from out in space. You will need to make several changes to the sky settings. Click on the Filters button, and select **Sun & Planets**. In the Filters window, make sure the fourth button (**Image**) for **Earth** is selected. Then click **OK**. Now click

7

on the large Location button, and select **Space: heliocentric** to switch the viewpoint out into space. Click on the **Lock Aim** button to unlock the Sun from the previous simulation. Now click on the small arrow to the right of the Aim box, choose **Planets/Moons**, and select **Earth** to rotate the sky to look at Earth. Now click on the **Lock Aim** button again to lock the Earth in the screen center. Then change the sky settings as shown in the following table.

Date	08/21/2017
Time	17:00
Step	1 hour
Location: Lon	150° 00′ 00″
Location: Lat	00° 00′ 00″
Location: Dist	1.000 au
Zoom	500

When you have changed all these settings, press the <Enter> key on the keyboard. You should see the Earth with the Moon just to its left. You are now looking at the Moon and Earth as seen from the direction of the Sun. Click two times on the Forward Step button to watch the first part of the eclipse again. Now that the Moon has just moved in front of the Earth, change the zoom to **9999**.

Q35 How large is the overall shadow compared with the diameter of the Earth? This is the approximate width of the Moon's outer shadow (penumbra) as it passes across the Earth's surface.

Finally, let's look at the eclipse as seen from the Moon. From the Events menu, again choose **Eclipses**. The August 21 eclipse should still be in the eclipse window. Make sure it is selected. If the **Change location to:** box is not checked, click on it to select it. Now click on the drop-down menu to the right of this box, and select **Moon surface**. Click **OK** in the Eclipse window. The Earth will reappear but in a rotated position. From the Control menu, select **Base Plane**, and click on **Ecliptic plane** to put the Earth back straight again. Now answer the following questions.

Q36 As you can see, the outer part of the Moon's shadow (the penumbra) covers a large area. However, most of that area will only see part of the Sun covered by the Moon. The actual total eclipse shadow (the umbra) is very narrow. Look for a second tiny dark shadow in the center of the large one. (If you have difficulty seeing the small shadow, click the

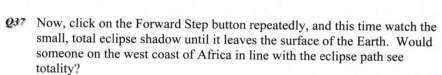

Forward or Reverse Step button until the eclipse shadow is over a cloud or lighter part of the map. Then it will be easier to see the small shadow.) This is the "total" part of the eclipse shadow. Estimate its size compared with the diameter of the Earth.

Q37 Now, click on the Forward Step button repeatedly, and this time watch the small, total eclipse shadow until it leaves the surface of the Earth. Would someone on the west coast of Africa in line with the eclipse path see totality?

Section 7.9

Select *RedShift 3/RedShift college edition* from the Windows menu to re-open the *RedShift College Edition* window. Click on the **Go** button in the navigation bar, select **Science of Astronomy**, then **Ancient Astronomers**, then **Prehistoric star gazers**. Finally, click on page **11** to go to the section on eclipses. Read, watch, and listen through page **19** of the Science of Astronomy. Now answer the following questions.

Q38 A solar eclipse can occur only at what phase of the Moon?

Q39 What phase must the Moon be in for a lunar eclipse to occur?

Q40 Why doesn't an eclipse occur every month?

Discussion

To the ancients, eclipses were dreadful events. These primitive people believed that some god was eating the Sun or turning the Moon to blood. Furthermore, eclipses did not happen often and were not easily predicted, so they seemed even more mystical. Today, eclipses are still interesting and sometimes awesome events, though we are less fearful. As you saw today, a given eclipse is only visible over part of the globe as the Earth's or Moon's shadow does not cover the whole planet. Since the Earth is larger than the Moon, it gets in the way of the Sun more often; thus, lunar eclipses are more common than solar ones.

Conclusion

In a couple of paragraphs, describe the conditions necessary for a lunar eclipse and for a solar eclipse. Tell how long they usually last. Explain why there isn't an eclipse each month even though the Moon orbits the Earth every 29 1/2 days. (Look back at your answer to Q40.)

Optional Observing Project

Do the observation exercise "The Moon" in Appendix C.

II
The Solar System

Contents

Name: _____

Course: _____

Section: _____

Retrograde Motion:
Ptolemy vs. Copernicus

Introduction and Purpose

Have you ever glanced up at the sky just after sunset and seen a really bright star you hadn't noticed before? Have you ever seen a bright star near the crescent moon that hadn't been there the night before? Most of the time, these "stars" that suddenly grab our attention are not stars at all, but planets. The ancients were fascinated by the planets because they were the only "stars" that moved around through the sky.

By plotting the planet motions, these ancient astronomers hoped to understand the shape and nature of our solar system. After many centuries of observing, two principal models for explaining planet motions emerged: the Ptolemaic model and the Copernican model. The Ptolemaic model placed the Earth at the center of our solar system, while the Copernican model put the Sun at the center.

The purpose of this lab is to help you begin building your own mental model of our solar system and to help you decide which of the two historic models best describes what we actually see. Do the following experiment using *RedShift*, and you will see the positions and paths of the planets from night to night just as those early astronomers saw them.

When you have followed the procedure and made your observations, we will ask you some questions about what you observed. When you are finished, you should have a better understanding of these two historic models of the Solar System.

Procedure

Section 8.1

Start *RedShift College Edition*, and then bring the *RedShift 3* screen to the front. In the Display menu, turn **Off** everything except the items marked **Sun & Planets** and **Stars**. In the Time settings panel, change the **Local** setting to **Universal**, and change the sky settings as shown in the table on the next page.

8

Universal date	06/01/2000
Step	10 days
Zoom	0.26

When all these changes have been made, press the <Enter> key on the keyboard. Now, select **Base Plane** from the Control menu, and click on **Eq. Earth** to fix our frame of reference. Click on the small arrow to the right of the Aim box, select **Stars**, click on **Polaris**, and then click the **Lock Aim** button to fix our viewpoint at the North Pole. This way, we can see the entire path of the planet as a circle as the planet moves all the way around the sky.

Section 8.2

Click on the Filters button, and select **Sun & Planets** to open the Sun & Planets filters window. Click on the large **Hide** button at the top to hide the Sun and all the planets. Now, click on the second button (**Icon**) of the object you wish to observe (begin with **Mercury**) to display the planet as a small colored symbol on the screen. Finally, click on the **Paths** button for this planet, and click **OK**.

Now change the **Universal** setting to **Julian date** in the Time settings panel. In the second column of the data table, write down the last five whole-number digits, omitting the decimal part of the number. (For example, 2450606.1174 would be recorded as 50606.) Move any open panels to the bottom of the screen so that you can see all of the sky.

Section 8.3

Place a piece of transparent tape on your monitor just below the initial position of the planet, and mark its starting point with a pen. Now you're ready to watch the motion of this planet over the months as it travels through the sky. When you run the simulation, watch for two different events: (1) the planet moving backward for a short time, and (2) the planet going all the way around the sky and returning to its starting point (marked on the tape). Whenever either of these things happens, note the Julian date and write it down in the data table, again recording only the last five digits and omitting the decimals.

Now click on the Forward Step button at the top of the Control Time panel to begin the simulation. Each time you click this button, you should see another symbol appear across the sky as each new position of the planet is plotted. Continue to click the Forward Step button, watching for the first reversal and/or the planet's return to its original point on the screen, and record the Julian dates accordingly.

After the first reversal is completed, continue clicking the Forward Step button to watch the planet's motion. If the planet passes its original point on the screen before the second reversal, be sure and record the Julian date when that occurs. Continue watching until a second reversal begins, note the Julian date, and write it down in the data table.

If the planet has not yet gone all the way across the screen and returned to its starting point, continue clicking until this happens. Once the planet has gone all the way around and back to its starting point and completed at least two reversals, you are done.

Section 8.4

Subtract the starting Julian date from the final Julian date. This is the *orbital period* of the motion of the planet relative to the Earth. Then subtract the Julian date of the beginning of the first reversal from the Julian date of the beginning of the second reversal to find the period between reversals (the *reversal period*). Record all this data in the data table on the next page.

Is the path smooth or does it contain a backwards loop or two? In the first column of the data table, write an **S** if it is smooth or an **L** if it has loops.

When you have completed recording your data for this planet, click on the Filters button, and select **Sun & Planets** to open the Filters window. Click on the **Hide** button for the currently selected object, and then click on the **Icon** button for the next object you wish to observe.

Reset the date to **06/01/2000**, and press the <Enter> key on your computer's keyboard. Now repeat Sections 8.3–8.4, selecting, in turn, Venus, Mars, Jupiter, and Saturn. Set the Time step parameters for each planet as shown in the table below.

Planet	Step
Venus	**10 days**
Mars	**15 days**
Jupiter	**40 days**
Saturn	**80 days**

8

Section 8.5

When the planet simulations have been completed, repeat the simulation using the Sun and then again using the Moon. (The Moon can be selected using the **Moons** tab at the left of the Filters window.) Set the time Step parameters for each of these objects as shown in the table below.

Object	Step
Sun	**15 days**
Moon	**2 days**

Watch each simulation until the Sun or Moon goes all the way around twice and comes back to its starting point the second time. Then record the final Julian date. Subtract the starting Julian date from the final Julian date and divide by two (since each went all the way around twice), and record this *orbital period* with respect to Earth in the data table. If neither the Sun nor the Moon reverses its motion during the simulation, we will assume that neither has a *reversal period*.

Data Table

Planet Motions							
Planet	Motion (S or L)	Start Date	1st Orbit Date	Orbital Period	1st Reversal Date	2nd Reversal Date	Reversal Period
Mercury							
Venus							
Mars							
Jupiter							
Saturn							
The Sun							
The Moon							

Section 8.6

Use the information in the data table to answer the following questions.

Q1 We know that the Moon orbits the Earth. Do any of the other objects move as though they were circling the Earth? (Are any of the other paths virtually the same as the Moon's?)

Q2 For the objects not listed in the previous question, which part of their motions would be difficult to explain if they were circling the Earth?

Q3   Arrange our solar system objects according to their orbital periods with respect to the Earth (from the shortest to the longest).
1)   2)   3)   4)
5)   6)   7)

Q4   Whirl something on the end of a keychain or string around your finger so that it winds around your finger until the string or chain is all wound up. Notice that the object goes around quicker and quicker as the string gets shorter and shorter.

If the Earth is at the center and the other objects are orbiting us like the keys on the keychain, which object should be the closest to the Earth? (Hint: Look back at Q3.)

Q5   Which should be farthest away?

Q6   Using a ruler and a sharp pencil, draw on the next page a scale model of the Solar System with the Earth at the center. As discussed in Q5, objects with a short period (the ones that orbit quickly) should be drawn closest to the Earth, and those with a long period (the slow ones) should be drawn farther out. To determine how big to draw each object's circle, use a scale of 30 days of orbital period per centimeter of radius. Use the following formula:

Radius of the circular orbit = **Orbital Period/30** *(days/cm)*

Here, for example, is the calculation for the Moon's circular orbit.
Radius of the Moon's orbit = **28** *(days)*/**30** *(days/cm)*
= **0.9 cm**
(Notice that we rounded off to 1 decimal place.)

Put the orbital periods from the data table into the formula above to calculate the radius of the circle you will draw for each of the objects listed, and record it below.

1) Mercury: 2) Venus: 3) Mars: 4) Jupiter:
5) Saturn: 6) Sun: 7) Moon: **0.9 cm**

8

In the space below, draw a dot in the center to represent the Earth. Using a compass or piece of string and a sharp pencil, draw a circle around the Earth with the proper radius (as calculated above) for each of the seven objects and label it. If an orbit is too big, just draw a circle outside the previous one, and label it with the name and correct diameter.

Ptolemy's Solar System

Q7 The picture you just drew is what Ptolemy's model looked like. Is this the order of the planets in the current model of the Solar System?

Q8 Now, arrange the objects according to their reversal periods (from shortest to longest).

1) 2) 3) 4)

5) 6) 7)

Q9 If the Solar System objects really are arranged as shown in your scale drawing with the Earth at the center, you would probably expect the closer ones to reverse more rapidly than the farther ones. Is that the case?

Q10 Can you see *any* way to predict the reversal periods using the orbital periods, or vice versa?

Section 8.7

As you have seen, the motion of each of the objects in the Solar System is different. The periodic backward motion of some of the objects in a loop or zigzag is called *retrograde motion*. Two of the objects, the Sun and the Moon, do not exhibit retrograde motion at all. Further, you saw that the retrograde motion did not seem to be related at all to the orbital motion.

Ptolemy tried to explain retrograde motion by suggesting that each planet rolled around slowly on the edge of a small wheel that was attached to the edge of a larger wheel that had the Earth at or near its center. Some planets had to have faster wheels, and some had to have slower wheels; some wheels had to be larger, and some smaller. Some of the objects with short orbital periods had long reversal periods, and others had short ones. There didn't seem to be any logical way to relate the orbital periods to the reversal periods. Now that you have observed and measured the paths of the planets, the Sun, and the Moon, you can appreciate the difficulty Ptolemy had in trying to build a single model that would correctly predict all their motions.

Over a thousand years later, Copernicus realized that if the Sun were at the center, and the Earth were revolving around the Sun along with all the other planets, such complexity would not be necessary. He proposed that each planet moved at a different speed. As we have already seen, closer planets would move faster, and farther ones would move slower. Suppose the Earth were somewhere out among the planets (instead of at the center of the Solar System). Then the combination of the motion of the Earth with the motion of each planet would explain the planet's backward loops

8

and its orbital period, as well as the continuous motion of the Sun. In the next few sections, we will see just how the Copernican Model works.

Section 8.8

Change the sky settings as shown in the table below, and press the keyboard <Enter> key.

Universal date	09/20/2001
Zoom	0.8

Click on the Filters button, and select **Sun & Planets** to open the Sun & Planets filters window. Click on the large **Hide** button at the top to hide the Sun and planets. Now click on **Jupiter's** second button (**Icon**), and make sure the **Paths** box is still checked. Click on the **Moons** tab at the left of the window to change to the Moons filters. Click the Moon's **Hide** button so the Moon won't be displayed, and then click **OK.**

Click on the small button to the right of the Aim box, choose **Constellations**, and select **Gemini** to move the sky around and bring Jupiter into view. Now, click on the **Lock Aim** button to lock Gemini in the center of the screen. From the Control menu, choose **Base Plane**, and click on **Ecliptic plane** to align the view screen with the ecliptic.

Finally, select **StopWatch** from the Tools menu to place the stopwatch to on your desktop. Set the stopwatch Step box to **10 days**. You are now ready to watch the motions of the planets Jupiter and Saturn. You will do Sections 8.9 and 8.10 for Jupiter and then repeat these sections for Saturn.

Section 8.9

Q11 When you click on the Forward Step button, the planet will move to the left for only a few steps before it stops and then begins a retrograde loop, moving backward to the right for a time.

Continue clicking until the planet stops moving to the right and resumes its normal motion to the left. Stop the motion exactly at this point. Use the Single Step forward and backward buttons to make sure that the planet is exactly at the stopping point in its retrograde motion when it is just about to resume its normal direction. Write down the **Elapsed time** shown on the stopwatch.

a)   Jupiter   b)   Saturn

Q12 Continue the simulation until the planet goes into retrograde motion a second time. (You may have to move the Control Panels to see it.) Let it move through this second retrograde loop until it is just at the point of stopping and resuming its normal motion to the left for the second time. Again, use the forward and backward Single Step buttons to make sure that the planet is just at the point of resuming its normal motion to the left. Write down the **Elapsed time** shown on the stopwatch.

a)   Jupiter   b)   Saturn

Q13 What was the period between retrograde loops for this planet? (Subtract your answer to Q11 from your answer to Q12.)

a)   Jupiter   b)   Saturn

Section 8.10

Now you are ready to repeat the simulation for Saturn. Click on the Filters button, and select **Sun & Planets** to open the Filters window. Click **Hide** in Jupiter's column and click **Icon** in Saturn's to turn Jupiter's icon **Off** and Saturn's **On**. Then click **OK** to close the Filters window. Change the date to **09/20/2002** and press the <Enter> key on the keyboard. Finally, click on the **Reset** button in the stopwatch to clear the stopwatch display. Now repeat Sections 8.9 and 8.10 with Saturn instead of Jupiter.

Section 8.11

Now we want to shift viewpoints to a position somewhere above the Sun and watch the same motion again. Click the Filters button and select **Sun & Planets** to open the Filters window. Now click on the second button (**Icons**) for the planets Earth and Jupiter and for the Sun. Click the **Hide** button for Saturn. (The **Paths** button should still be selected.) Then click **OK** to close the Planets filters window. Click the small Close button at the upper right corner of the stopwatch tool to get it off the screen. Finally, select **Fine Grid Overlay** from the Tools menu to put the grid on the screen.

In the Location panel, click on the large location button, and change it to **Space: heliocentric**. If the Lon (longitude) and Lat (latitude) boxes are not visible in the Location panel, click on the small icon at the upper right corner of the Location panel ⌑ to expand it. Now change the Location panel values as shown in the following table, and press the keyboard <Enter> key.

8

Lon (longitude)	**270° 00′ 00″**
Lat (latitude)	**90° 00′ 00″**
Dist (distance)	**20 au**

This setting places you out in space 20 Astronomical Units, directly above the North Pole of the Sun, looking down on the Solar System. Now click on the **Lock Aim** button to lock the Sun in the screen center, change the sky settings as shown in the table below, and press the <Enter> key on your keyboard.

Universal date	**09/20/2001**
Step	**1461 hours**
Zoom	**1**

Now you should see the Sun at the center of the screen with Earth to the right at the 3:00 position and Jupiter directly above the Sun. You now will do Section 8.12 for Jupiter and then repeat it for Saturn.

Section 8.12

Click repeatedly on the Forward Step button in the Control Time panel, and watch the simulation through 12 steps (2 years). (The Earth's positions will overlap for the second 6 updates.) Now, from the File menu, choose **Print** to print a copy of your screen.

Once you have your printout, look at Jupiter's motion. Does it show any retrograde motion? To see how Jupiter's *apparent* retrograde motion as seen from the Earth is actually caused, go through the following procedure.

Look at the example illustration on the next page. It is a plot of the positions of Mars, the Sun, and Earth. You will make a similar one for Jupiter (and then Saturn). First, draw a line on your printout from the first position of the Earth (the one at about a 1 o'clock position) to the first position of Jupiter, and label it "1". Then draw a line from the second position of the Earth (the 11 o'clock one) to the second position of Jupiter, and label it "2". Continue through the first six Earth positions, drawing the lines and labeling them. *At this point, Earth has completed one year, and its positions begin to repeat.*

Continue to draw lines from the Earth's 7–12 positions (repeating its 1–6 positions) to Jupiter's 7–12 positions, labeling each one as you go. Follow the example below, which uses the motion of Mars instead of Jupiter.

Each of these lines represents the *apparent* position of the planet as seen from the Earth, measured with respect to the background stars (the edges of the paper in our experiment).

When you are finished drawing the lines and labeling them, draw a line connecting the ends of the 12 lines *in order* as shown in the example below. This will create a plot of the path of Jupiter through the sky *as we see it from the Earth*. As you can see below, the *apparent* motion of Mars reverses direction for a couple of months—retrograde motion! The motion of Jupiter on your printout should do the same. (See below for Saturn.)

Sample Data Interpretation

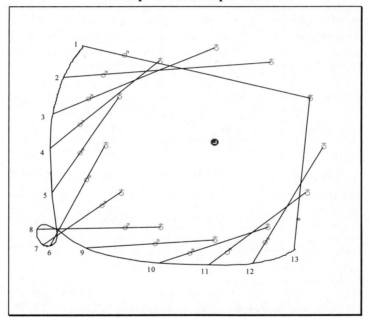

Mars's apparent motion as seen from Earth

Section 8.13

Now we want to watch Saturn instead of Jupiter. Open the Sun & Planets filters window, and click in the **Hide** column for Jupiter to hide it. Click in the **Icon** column next to Saturn to show it as an icon, and click **OK**.

8

Change the sky settings as shown in the following table, and then press the <Enter> key on your computer's keyboard.

Universal date	09/20/2002
Dist (Location Panel)	**40 au**

Now, Saturn should be above the Sun with the Earth again at the 3:00 position. Repeat Section 8.12, and do the simulation again for Saturn.

Section 8.14

When you have completed printing out and plotting the retrograde motions of both Jupiter and Saturn, answer the following questions.

Q14 On your printouts of the planets' motions, count the number of intervals between the end of the first retrograde loop and the end of the second retrograde loop for Jupiter and then for Saturn, and write these answers below:
1) Jupiter   2) Saturn

Q15 Since each of these intervals was approximately 61 days, multiply your answers to Q14 by 61, and write those numbers in below.
1) Jupiter   **(Q14)** × 61 = (   ) days
2) Saturn   **(Q14)** × 61 = (   ) days

Q16 How do your answers to Q15 compare with your answers to Q13?

Q17 Give a reason why your answers in Q15 might not be exactly the same as in Q13.

Discussion

In today's exercise, we observed the paths of the planets, Sun, and Moon as seen from Earth. We saw that their motions were difficult to predict from an Earth-based point of view. Some of the objects displayed retrograde motion, and others did not.

Then we looked at the motions of Jupiter and Saturn from the heliocentric viewpoint. We found that the heliocentric model of Copernicus, combining the motion of the Earth and the motion of these planets, described accurately what we saw from Earth.

The inner planets can also be modeled in the same way. However, because these planets go around the Sun much faster than Earth, it is a little more difficult to tell which position is which on the printouts. If you do the simulation, though, you will find that the results there are just as accurate as those that we did for the outer planets.

Conclusion

In a few sentences on the next page, describe the motion of the planets, Sun, and Moon as seen from Earth. In your summary, explain what retrograde motion is. Describe Ptolemy's model and comment on its difficulties. Then give a brief description of Copernicus' model and tell how it explained retrograde motion. Finally, give some reasons for preferring either Copernicus' model or Ptolemy's model. (Do not include data from modern spacecraft as a part of your reasoning.)

When faced with more than one potentially correct explanation for an observation, we often apply a principle called ***Occam's Razor***. Occam's Razor states that when two explanations are possible, the one with the fewest assumptions, complexities, and difficulties is probably the best explanation. Write a short conclusion (three or four sentences) summarizing what you have learned about the motion of the planets as viewed from Earth, and (using Occam's Razor) discuss which model, the Ptolemaic or the Copernican, is probably the better model of the Solar System.

Optional Observing Project

Do the observation exercise "The Planets" in Appendix C.

8

Notes

Name: _____

Course: _____

Section: _____

Kepler's Third Law

Introduction and Purpose

In 1600, Tycho Brahe and Johannes Kepler were the first to study the motion of the planets using a systematic, mathematical approach. Brahe was a Danish nobleman who used the most precise instruments available to plot the positions of the planets as they moved through the stars. Brahe hired Kepler, a mathematician, to fit his data into a set of mathematical equations.

Kepler worked for many years trying to make the planet positions fit onto circular paths, but circles just didn't work. Then he tried to find orbits based on other perfect mathematical figures such as pentagons, hexagons, octagons, etc., but to no avail. Finally, Kepler stumbled upon the correct form of equation for the orbits—an ellipse.

Combining Brahe's data with Copernicus's heliocentric model, Kepler was able to deduce the shape of the planet orbits, their relative distances from the Sun, and their periods. In this process, he discovered three laws that governed every planet for which he had data. His third law seemed particularly strange. He found that if you cubed the orbital radius of a planet and divided it by the orbital period squared, the number (R^3/P^2) was the same for every planet!

To understand the amazing nature of Kepler's work, you must realize that in Kepler's day, the only view available was from Earth. He wasn't even certain that the planets orbited the Sun. In today's exercise we will put ourselves in Kepler's and Brahe's place. We will make the same kind of observations Brahe made and use the same mathematical logic Kepler used. We will see how they measured the time it took for each planet to go around the Sun (its period) and how they deduced each planet's relative distance from the Sun.

RedShift, however, will still give us many advantages they did not have. Brahe had to observe at all hours of the night to see all parts of the sky; we can shrink the sky to see it all at once. Brahe had to observe some of the planets for many years to see one complete orbit; we can watch the same motion in a few minutes. With *RedShift's* help, we will use the heliocentric model of the Solar System to deduce the information Kepler deduced. We will deduce the period of the planets as they orbit around the Sun and the relative distance of each planet from the Sun. Finally, we will use that information to show that Kepler's third law holds true.

Procedure

Section 9.1

9

Start *RedShift College Edition*, and then bring the *RedShift 3* screen to the front. In the Display menu, turn **Off** everything *except* the items marked **Planets** and **Stars**. Click on the small arrow to the right of the Aim box, and select **Sun** to center the screen on the Sun. Then click on the **Lock Aim** button to lock the Sun in the center. Now answer the following questions to get some important background information.

Q1 How many days does it take the Earth to go around the Sun once? (How many days are in a year?)

Q2 If the Earth travels one full circle in its orbit in one year (365.25 days), what fraction of an orbit does the Earth travel in one day? (Divide 1 by the number of days.)

Orbits/day = 1 orbit/365.25
Orbits/day = 1 orbit/ 365.25
= (   ) Orbits/day

(This is a *very* critical number. Record it accurate to 7 decimal places, both here and in the data table.)

Section 9.2

First, we will measure the orbital periods of two of the five naked-eye planets. (We could measure them all, but it takes a lot of time.) The process Kepler and Brahe used to measure the planets inside the Earth's orbit is different from the process used for the planets outside the Earth's orbit. So we will choose one of the inner planets, Mercury, and one of the outer planets, Jupiter, to let you see how they did it.

The measurement of orbital period for an inner planet depended on measuring the planet's *conjunctions* with the Sun. Whenever two celestial objects move past one another in the sky, it is called a *conjunction*. In the next section, you will observe and measure three successive conjunctions of Mercury with the Sun and then see how that helps us establish Mercury's orbital period.

From the Control menu, select **Base Plane**, and click on **Eq. Earth** to set our viewpoint. Click on the Filters button, and select **Sun & Planets** to open the Filters window. Click on the **Hide** button at the top of the window to hide all the planets, and then click in the second column (**Icon**) after **Mercury** and the **Sun**. Then click **OK**. From the Tools menu, select **StopWatch** to open the stopwatch tool.

Section 9.3

In the Time Settings panel, change the **Local** setting to **Universal**, change the sky settings as shown in the following table, and press the <Enter> key on the keyboard.

Universal date	**07/06/2000**
Universal time	**4:00**

Now change the Stopwatch Step as shown in the next table.

Stopwatch Step	**2 days**

Now you should see the Sun in the center of the screen and Mercury's symbol directly below the Sun. This is the first *conjunction* of our experiment. Now click on the **Reset** button in the Stopwatch, and answer the following questions.

Q3 Click on the Forward Step button repeatedly. The planet will go to one side, stop, go back past the Sun (a second conjunction), and then go to the other side. Continue until the planet is again lined up with the Sun (the third conjunction).

Change the Stopwatch Step to **1 day**, and use the Single Step buttons to position the planet as close to the Sun as possible. Record the Elapsed time shown in the Stopwatch in the data table's first column.

Q4 If the Earth was stationary, the time you just measured would be Mercury's period. However, during this simulation, the Earth also traveled along *its* orbit, so the total time was actually more than one orbit of Mercury. To see why this is true, you must imagine what these three conjunctions would have looked like from out in space. Select *RedShift 3/RedShift* **college edition** from the Window menu to bring the *RedShift College Edition* screen to the front. Click on the **Go** button in the navigation bar, then select **Principia Mathematica** and **Kepler's Laws** and click on page **10**. Here you will see the same simulation you just observed from a viewpoint in space. Remember, Kepler could only imagine what this was like. Watch the simulation. (Click the **Replay** button if you want to watch the simulation again.)

9

In order to find the actual time it took Mercury to go around the Sun once, we must find how much of an extra orbit it traveled during the simulation and then reduce the total time by a similar fraction.

As you saw in the simulation, the extra fraction of Mercury's orbit will be the same as the fraction of an orbit traveled by Earth. Since each day of Earth's travel corresponds to a tiny fraction of an orbit (see Q2), we can calculate the total fraction traveled during the simulation. What fraction of an orbit did the Earth complete in *its* travel during this time? [Multiply the number of days you calculated in Q3 by the number of orbits per day the Earth travels (from Q2).]

Fraction of an orbit = Number of days × Number of orbits/day
$$= \textbf{(Q3)} \times \textbf{(Q2)}$$
$$= (\quad\quad) \times (\quad\quad\quad)$$
$$= (\quad\quad\quad)$$

Round this answer to 4 decimal places, and write it in the second column of the data table.

Section 9.4

Now, let's shift the viewpoint away from Earth to a position above the Sun so you can do the simulation for yourself. (This was something Kepler could only imagine.) Bring the *RedShift 3* screen back to the front. In the Location panel, click on the large Location button, and select **Space: heliocentric**. If the Lon (longitude) and Lat (latitude) boxes are not visible in the Location panel, click on the small icon at the upper right corner of the Location panel to expand it. Click on the Filters button, and select **Sun & Planets**. In the Filters window, click on the second button (**Icon**) after **Earth** to add it to the display, and then click **OK**.

Change the sky settings and location coordinates as shown in the table below and press the <Enter> key on the computer keyboard.

Universal date	07/06/2000
Universal time	4:00
Lon (longitude)	180° 00′ 00″
Lat (latitude)	90° 00′ 00″
Dist (distance)	4 au
Stopwatch Step	2 days

This will place you 2 astronomical units above the Sun's North Pole, looking down on the Solar System. You should see the Sun, the Earth, and Mercury all lined up. Now answer the next question.

Q5   Count how many orbits Mercury makes between the first and third conjunctions. (Remember, you are starting at the first conjunction.) Click on the Forward Step button in the Control Time panel, and watch the simulation again through the third conjunction. *(Be sure to count the conjunction when Mercury is on the opposite side of the Sun from the Earth as well as when it is on the same side. If you're not sure how to do this correctly, continue the simulation until the elapsed time shown in the stopwatch is approximately the same as the time you measured in watching the simulation from Earth.)*

How many *complete* orbits (ignoring any fractional parts) did Mercury make during this simulation?

Section 9.5

The planet's orbital period, **P**, is the length of time required to go around the Sun one time. To calculate the period of Mercury, we need to know two things: the length of time between the first and third conjunctions (which we have just measured) and the total number of orbits Mercury made during that time. The period will be the total length of time divided by the total number of orbits.

Q6   Now, we need to calculate the *exact* number of orbits (including fractional parts) Mercury made during that time.
Total number of orbits traveled by Mercury = 1 + (**Q4**)
$$= (\qquad)$$

Write this answer in the third column of the data table.

Q7   Finally, we are ready to calculate Mercury's orbital period, **P**. We will first divide the number of days between conjunctions (the first column of the data table) by the number of orbits Mercury made during that time (the third column of the data table). This is the period in days.

$$\text{Period (}\mathbf{P}\text{)} = \frac{\text{Days between conjunctions}}{\text{Total number of orbits}}$$

$$= \frac{\mathbf{1st\ column}}{\mathbf{3rd\ column}} = \frac{(\qquad)}{(\qquad)} = (\qquad) \text{ days}$$

Round off the answer to two decimal places, and record it in the fourth column in the data table.

9

Section 9.6

Now let's move to the planets beyond the Earth. To measure the orbital period of these planets, Kepler used Brahe's observations of successive times of *opposition*. *Opposition* is the point where the planet is exactly opposite the Sun in the sky—just rising as the Sun is setting.

When the planet is just rising as the Sun is setting, it is at opposition, and the planet, Earth, and Sun are all lined up just as they are at conjunction. For the outer planets, we will measure the time between two successive oppositions rather than between the first and third conjunctions. You will do this experiment for Jupiter.

Section 9.7

Click on the small arrow to the right of the top Location box (which says **Heliocentric**), choose **Planets/Moons**, and select **Earth** to change the location back to the surface of the Earth.

Open the **Sun & Planets** filters window, and **Hide Mercury** and **Earth**. Click the **Icon** button for **Jupiter**, and then click **OK**. Now change the sky and location settings as shown in the following table, and press the <Enter> key on your computer keyboard.

Universal date	06/01/1995
Universal time	17:57
Lon (longitude)	000° 00′ 00″
Lat (latitude)	00° 00′ 00″
Zoom	0.5
Stopwatch Step	10 Days

These settings put us at Earth's equator. Open the Guides filters window, change the **Horizon** setting to **Line**, and click **OK**. Now, from the Display menu, click on **Guides** to turn on the horizon line. Click on the small arrow to the right of the Aim box, and select **Sun** to rotate the sky to center on the Sun. (**Unlock** the Sun if it is locked.) It should just be setting on the western horizon. To verify that this is a point of opposition, again click on the small arrow to the right of the Aim box. This time, choose **Planets/Moons**, and then click on **Jupiter**. Now you should see Jupiter's symbol at the center of the screen just rising on the eastern horizon. Click on the **Reset** button in the Stopwatch tool to set the display to **0**.

Q8   Click on the Forward Step button repeatedly. Jupiter will move up to the top of the screen and disappear. Just as Brahe might have done, continue to watch until Jupiter reappears (at the bottom of the screen) moving toward the horizon. Just when Jupiter has almost reached the horizon, change the Stopwatch Step to **1 day**, and continue until the eastern horizon line goes right through the middle of the planet's symbol.

This will not be the exact date of opposition. Use the Aim box, and select **Sun** to check the position of the Sun. Click in the Time Settings box, and adjust the time a few minutes earlier or later so that the western horizon line goes right through the Sun.

After you change the time, use the Aim box again and reselect **Jupiter** to see where it is now. If it has shifted, move a day or two forward or backward to get Jupiter right on the horizon line again. Finally, use the Aim box to recenter the screen on the Sun.

Repeat these steps until you have found the exact date and time when the western horizon line goes right through the Sun and the eastern horizon line goes right through Jupiter. When you have gotten as close to opposition as possible, record the **Elapsed time** shown on the Stopwatch here and in the first column of the data table.

   Elapsed time between oppositions = (   )

Q9   As before, we wish to compute the number of orbits the Earth completed during this simulation. (Multiply the number of days recorded in Q8 by the number of orbits/day calculated in Q2.)
   Number of orbits = Number of days × Number of orbits/day
   = (Q8) × (Q2)
   = (   ) × (   ) = (   )

Round the value to 4 decimal places, and write it in the second column of the data table.

Section 9.8

Now, as before, we want to shift our viewpoint to a position above the Sun and repeat the simulation. Click on the large Location button in the Location panel, select **Space: heliocentric**, and then **lock** the Sun in the screen center. Click on the Filters button, and select **Sun & Planets**. In the Filters window, click on the second button (**Icon**) after **Earth** to add it to the display, and then click **OK**.

111

9

Now change the sky and location settings as shown in the following table, and press the <Enter> key on your computer keyboard.

Universal date	06/01/1995
Universal time	17:57
Lon (longitude)	000° 00′ 00″
Lat (latitude)	90° 00′ 00″
Dist (distance)	6 au
Stopwatch Step	10 days

You should now see the Sun, the Earth, and Jupiter all in a line. (You may need to move and **Reset** the Stopwatch to see everything.)

Q10 Click on the Forward Step button in the Control Time panel, and watch the simulation until the Earth catches back up to Jupiter and the Sun, Earth, and Jupiter are again lined up (the second opposition). How many complete orbits (ignoring fractional parts) did Jupiter make during this simulation? (If Jupiter did not complete an orbit, write **0**.)

Section 9.9

For Jupiter, the situation is reversed from Mercury. As you saw, the Earth traveled *more* than one complete orbit, while Jupiter only traveled a small fraction of an orbit. Again, however, the fractional parts of each orbit are the same. This time, the fractional part of Jupiter's orbit will be the same as the fractional part of Earth's travel beyond the first orbit. To see why this is true, again select *RedShift 3/RedShift* **college edition** from the Window menu to bring the *RedShift College Edition* screen to the front. Click the **Go** button in the navigation bar, select **Principia Mathematica** and **Kepler's Laws**, and click on page **18**. Watch the simulation. (Click the **Replay** button if you want to watch the simulation again.)

Q11 The exact number of orbits traveled by Jupiter is just the *decimal fraction* of Earth's travel beyond one complete orbit (from Q9). Write this number in the third column of the data table.

Orbits traveled by Jupiter = **(only the decimal part of Q9)**
= (   ) orbits

Q12 Once again, we are ready to calculate Jupiter's orbital period, **P**. As before, divide the number of days between oppositions (the first column of the data table) by the number of orbits Jupiter made during that time (the third column of the data table). This is the period in days.

$$\text{Period (P)} = \frac{\text{Days}}{\text{Number of orbits}}$$

$$= \frac{\text{1st column}}{\text{3rd column}} = \frac{(\qquad)}{(\qquad)} = (\qquad) \text{ days}$$

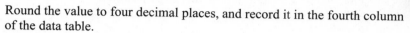

Round the value to four decimal places, and record it in the fourth column of the data table.

Section 9.10

To make the numbers easier to manage, you need to convert these orbital periods from days into years. To do this, divide the orbital period of each planet (column three) by **365.25** (the number of days in one year). After dividing, record the answer in the fifth column of the data table. Repeat this procedure for each of the five planets.

Data Table

Part of an orbit traveled by the Earth per day (from Section 9.1)	

Orbital Data for the Naked-eye Planets								
Planet Name	Elapsed Time (days)	Number of Orbits (Earth)	Number of Orbits (Planet)	Orbital Period, P (days)	Orbital Period, P (years)	Angle of Maximum Elongation	Orbital Radius, R (au)	$\frac{R^3}{P^2}$
Mercury								
Venus				225	0.62		0.72	
Mars				687	1.88		1.52	
Jupiter							5.20	
Saturn				10,746	29.42		9.54	

*Note: This exercise involves many different measurements and computations. While none of the measurements or computations is very difficult, a small mistake can easily throw off the results. This same experiment can be found in the **Principia Mathematica** segment titled **Kepler's Laws** in RedShift College Edition.*

*If you have difficulty following the instructions here, or if your results are not consistent, go to the Window menu and select **RedShift 3/RedShift college edition**. In the RedShift College Edition window, click on the **Go** button in the navigation bar, select **Principia Mathematica** and **Kepler's Laws**, and then click on page **9**. Follow the text and perform the experiments through page 25. Then come back here and try this exercise again.*

9

Section 9.11

Now that you have measured the orbital periods of two of the planets as Kepler did, we want you to see how Kepler deduced the relative orbital radii. This is fairly straightforward for the *inferior* planets, and we will show you how to do it for Mercury. The procedure is more complicated for the *superior* planets, so we have omitted it from this exercise.

You can measure the radius of the orbit of an inner planet by observing the points where it is the farthest from the Sun—the points of *maximum elongation*. In the following drawing, you can see the geometry of Mercury's position at maximum elongation.

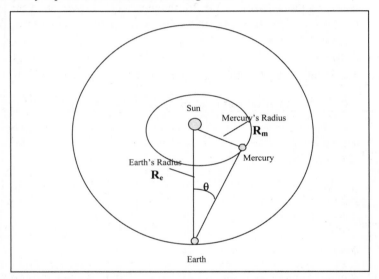

The radius of Mercury (relative to Earth), R_m, can be calculated by measuring the angle of maximum elongation, θ, and using this formula:

$$R_m = R_e \sin \theta.$$

As you can see from the diagram, when the orbit of a planet is not quite circular, we must take two measurements of the angle of elongation, one on each side of the Sun, to get a more accurate average value for **R**.

Section 9.12

To measure the relative orbital radius of Mercury's orbit, bring the *RedShift 3* screen back to the front, and close the **StopWatch** tool. Re-open the Sun & Planets filters window, hide **Jupiter**, click on the **Icon** button for **Mercury**, and click **OK**. Now click on the small arrow at the right of the top location box (which is set to **heliocentric**), select **Planets/Moons**, and click on **Earth** to change the location back to the surface of the Earth. Change the sky settings as shown in the following table, and press the <Enter> key on the computer keyboard.

Universal date	07/06/2000
Universal time	12:00
Step	1 day

Click on small arrow at the right of the Aim box, and select **Sun** to center the screen on the Sun. Again, use the **Lock Aim** button to lock the sun in the center. You should now see the Sun centered on the screen with Mercury nearby. Now select **Ruler** from the Tools menu to place the Ruler tool on your screen. The computer is now set to measure the relative orbital radius of Mercury.

Q13 Click on the Forward Step button repeatedly. Watch until Mercury reaches its most distant point and begins to move back toward the Sun. Stop here, change the Step to **6 hours**, and then move forward or backward until you reach the point at which Mercury is exactly at its greatest distance from the Sun—maximum elongation.

Using the Ruler tool, click first on the center of the sun and then click again on Mercury. In the space below, record the number of degrees shown in the **Result angle** box of the Ruler tool.

Mercury's elongation (1) = (   ) degrees

Round your answer off to 2 decimal places.

Q14 Change the Step as shown and press the <Enter> key on the computer keyboard.

Step	1 day

Now click repeatedly on the Reverse Step button until Mercury reaches its farthest point from the Sun in the opposite direction. As before, change the Step to **6 hours**, and then move forward or backward until you reach the point at which Mercury is exactly at its greatest distance from the Sun—maximum elongation.

Again use the ruler tool to measure this maximum elongation, and record the value shown in the **Result angle** box in the appropriate space below.

9

Mercury's elongation (2) = (   ) degrees

Round your answer off to 2 decimal places.

Q15 We now have the maximum elongation to the east and to the west for Mercury and need to find the average value. Add the two values together and divide by two, and then record your answer in the **Angle of Maximum Elongation** column of the data table.

$$\frac{Q13 + Q14}{2} = \text{Average elongation}$$

Mercury:

$$= \frac{(\qquad) + (\qquad)}{2} = (\qquad) \text{ degrees}$$

Round your answer off to 2 decimal places.

Q16 Now to find the actual radius of the orbit in terms of Earth's radius (AU), put the average value, Q15, into the equation given earlier: $R_m = R_e \sin \theta$.

Mercury: $R_m = R_e \sin$ **(Q15)**

                  $= (1 \text{ AU}) \sin(\qquad)$

                  $= (\qquad) \text{ AU}$

(If you don't have a calculator that will do sines or don't know how to do them, get your instructor to help you.)

Round this answer off to 2 decimal places, and record it in the **Orbital Radius** column of the data table.

Section 9.13

Now you have measured the orbital periods of two of the planets (and the radius of one). We have supplied these values for the rest of the five naked-eye planets, and it is time to put Kepler's third law to the test. Following the steps in the example below, calculate the ratio of R^3 / P^2 for each of the five planets. Round the ratios off to 2 decimal places, and write them in the last column of the data table.

Example 1: Using Earth's radius (1 au) and period (1 year).

E1   Calculate R^3. $R^3 = (R)(R)(R)$

                           $= (1 \text{ au})(1 \text{ au})(1 \text{ au})$

                           $= (1) \text{ au}^3$

E2   Calculate P^2. $P^2 = (P)(P)$

$$= (1\ year)(1\ year)$$
$$= (1)\ year^2$$

E3   Divide $\mathbf{R}^3$ by $\mathbf{P}^2$. $\dfrac{\mathbf{R}^3}{\mathbf{P}^2} = \dfrac{(E1)}{(E2)}$

$$= \dfrac{(1\ au^3)}{(1\ year^2)}$$

$$= 1\ au^3/year^2$$

Q17   After having computed the ratio of R^3 to P^2 for all five of the planets, calculate the average value of $\mathbf{R}^3/\mathbf{P}^2$. (Add all the values in the last column of the data table and divide by 5.) Record this value here.

Q18   Which planet's data are the farthest off from this average value?

Q19   How far off are they? Subtract the planet's value from the average value (Q17). (If the result comes out negative, drop the minus sign—this makes it an *absolute value*.)

Q20   Scientists like to express the accuracy of a theory by comparing the values predicted by the theory with different measured values. This comparison is a quantity known as *percent uncertainty* (or sometimes as *percent error*). The maximum *percent uncertainty* of a measurement is found by subtracting the worst measured value from the average value, taking the *absolute value*, and then dividing this difference by the average value and multiplying by 100. Here is the formula:

Maximum % uncertainty $= \left| \dfrac{(Average\ Value)\ \text{-}\ (Worst\ Measured\ Value)}{(Average\ Value)} \right| * 100$

What is the percent uncertainty of your worst measurement?

% uncertainty $= \dfrac{(Q19)}{(Q17)} * 100$

$$= \left(\dfrac{\quad}{\quad} \right) * 100$$

$$= (\quad)$$

Round your answer off to 1 decimal place.

Q21   The experimental uncertainty in this experiment was affected by many factors. Some of these factors were your ability to measure the distances

117

9

from the printouts or screen, the precision with which the points were plotted, and the precision with which you were able to determine the exact dates of conjunction or maximum elongation. Any error in any of these measurements would be compounded twice in squaring the period, and three more times in cubing the radius. So even an error of 3% could result in an overall uncertainty of 15%.

Is your answer to Q20 less than 15%? If so, you can confidently say that Kepler's third law holds true for the five naked-eye planets.

Q22 So, if the ratio of R^3/P^2 is the same for every planet, use the average value you computed for R^3/P^2, and calculate the period of a planet that had an orbital radius of **39.4 au** (the orbital radius of Pluto).

Discussion

Kepler's work with Brahe's observations proved pivotal to advancing our understanding of the shapes of the orbits and the nature of the motions of the bodies in our solar system. Isaac Newton used Kepler's mathematical analysis of the motion of the planets and their orbital radii to confirm his own suspicions about the nature of gravity. And Edmund Halley used Kepler's *and* Newton's theories to analyze the orbits of a few comets. As a result, he successfully predicted the reappearance of the comet that has ever since been called by his name.

As you can imagine, Kepler's and Brahe's work was much more tedious without the aid of mathematical modeling machines and printouts like those we used today. What we did in a few minutes took them years of exacting observation and mathematical trial and error. But it involved nothing you could not do today with the simplest of instruments if you spent the time necessary to do it.

Conclusion

On a separate sheet of paper, describe the process of measuring the orbital periods of the planets and their orbital radii. In looking at your data table, what can you say about the relationship between the period of a planet's motion and the distance at which it orbits? Write a rule to express that relationship mathematically.

Name: _____

Course: _____

Section: _____

Galilean Merry-Go-Round

Introduction and Purpose

An excellent test of Newton's theory of gravity came in observing the four bright Galilean moons of Jupiter. Because these moons are so bright and far from the planet, they can be observed even with small telescopes and binoculars. After observing these four moons over time, Galileo realized that here was a miniature of our solar system. As was true of the Keplerian planet motions, the closer moons moved faster and the farther ones moved slower.

Once Isaac Newton had constructed his model of gravity, the motion of these moons could easily be explained. A very exciting result of Newton's laws was that if the orbital period of a moon and its distance from the planet could be determined, you could deduce the mass of the planet. In other words, we could figure out just how heavy Jupiter was by watching its moons!

The periods of the moons were easy to measure. Once the distance to Jupiter had been measured, the distance from each moon to Jupiter (the orbital radius) could also be determined. The purpose of today's exercise is to determine the mass of Jupiter. First, we will measure the periods of the moons by observing their positions night after night for a period of time. We could also measure their orbital radii, but *RedShift* does not have sufficiently fine grids to do this easily, so those values have already been entered into your data table. Finally, we will plug these two quantities into Newton's laws of motion and gravity and deduce the mass of Jupiter. We will repeat the process for each of the four bright moons of Jupiter and thus get four independent values for Jupiter's mass. Then we will compare our results and see just how well this relationship works.

Procedure

Section 10.1

Start *RedShift College Edition*, and then bring the *RedShift 3* screen to the front. When the *RedShift 3* screen opens, click on the File menu, and select **Open Settings** to open your personal settings file. In the Display menu, turn off all items except **Planets**, **Moons**, and **Stars**.

Click on the Filters button and select **Moons** to open the Moons filters window. In this window, select **Jupiter** in the Planet: box, and click the large **Hide** button at the top of the window to hide all of Jupiter's moons.

10

Now click in the fourth column (**Image**) of the moons **Io**, **Europa**, **Ganymede**, and **Callisto** to make sure that these moons show on the screen. Then click **OK**. Finally, change the sky settings as shown in the table below, and press the keyboard <Enter> key.

Date	Today's Date
Local Time	9:30 p.m.
Step	1 Day
Zoom	20

Click on the small arrow to the right of the Aim box, select **Planets/Moons** and then click on **Jupiter** to rotate the screen to center on Jupiter. Now click the **Lock Aim** button to lock the screen on Jupiter.

This view of Jupiter is very similar to what you would see with a small telescope or large binoculars. Jupiter should look like a small, round disc in the center with several moons arrayed in a straight line (more or less) on either side of the planet. It was this straight-line arrangement that caught Galileo's attention. Sketch this view in the space below.

Section 10.2

In the Control menu, select **Base Plane**, and click on **Ecliptic plane**. This setting will rotate the screen to line up (more or less) with the plane of Jupiter's moons. To see what Galileo saw as he first observed Jupiter over a period of time, click repeatedly on the Forward Step button. Is the arrangement of the moons ever the same on any two consecutive nights? Using only a good pair of binoculars, you can make the same observations Galileo did.

Again, open the Moons filters window. Jupiter should still be selected in the planet box. Now Hide all the moons except **Io**, and click **OK**. Change the sky settings as shown in the table below, and press the keyboard <Enter> key.

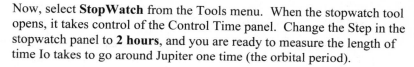

Date	Today's Date
Local Time	9:30 p.m.
Zoom	100

Now, select **StopWatch** from the Tools menu. When the stopwatch tool opens, it takes control of the Control Time panel. Change the Step in the stopwatch panel to **2 hours**, and you are ready to measure the length of time Io takes to go around Jupiter one time (the orbital period).

Section 10.3

Repeatedly click the Forward Step button in the Control Time panel until the moon (**Io**, to begin with) is at its farthest point to the right of Jupiter. (Mark this point with transparent tape on your monitor.) Then click on the **Reset** button in the Stopwatch tool to set its display to **0**.

Now click on the Forward Step button repeatedly until the moon (Io, to begin with) has gone all the way across to the left side of Jupiter and then all the way back to the right. This is one complete orbit. Continue the simulation through three complete orbits. At the end, make sure you are as close to the point of greatest distance as possible. In the space below, record the Elapsed time shown on the Stopwatch display. This value is the number of days required for three complete orbits of the moon. To get the period for one orbit, divide this number by 3, round it off to two decimal places, and write it in the data table column titled "Orbital Period."

Section 10.4

To measure the orbital periods of Europa, Ganymede, and Callisto, we will change the Stopwatch Step parameter as shown in the table below.

Moon	Stopwatch Step
Europa	6 hours
Ganymede	6 hours
Callisto	12 hours

Open the Moons filters window again, Hide the previous moon (Io, to begin with), click on the fourth button (**Image**) of the next moon in the

10

list, and click **OK**. Repeat this section to obtain the orbital periods for each of the other three moons.

Section 10.5

Before we can calculate the mass of Jupiter, we need to know Newton's equation for making that calculation. The equation is:

$$M_{jupiter} = (4\pi^2/G) * (r^3/P^2)$$

The first part of this equation, $(4\pi^2/G)$, always has the same value. (It is a *constant*.) Its value is $\mathbf{7.9 \times 10^{10}}$ kg * days^2/km^3.

The second part of the equation, (r^3/P^2), is a statement of Kepler's third law that we studied in a previous exercise. **R** is the radius of the moon's orbit around Jupiter, and **P** is the period of one orbit. Kepler's third law says that the ratio of $\mathbf{r^3}$ to $\mathbf{P^2}$ is a constant for all objects orbiting a given mass (such as the Sun or Jupiter).

In today's exercise, you have already measured the orbital period (**P**) for each of the moons and have been given the values for the radius of each moon's orbit, (**r**). To make our final calculations easier, let's begin by squaring the period and cubing the radius for each moon. Use your calculator to square each of the Period values, and enter the results in the $\mathbf{P^2}$ column of the data table. Then cube the radius of orbit (r) [multiply **r** * **r** * **r**], and enter this value in the $\mathbf{r^3}$ column of the data table.

To calculate the mass of Jupiter, multiply the constant, $\mathbf{7.9 \times 10^{10}}$ kg * days^2/km^3 times the r^3 column, and divide by the P^2 column. Then write this value, the mass of Jupiter, in the Mass of Jupiter column. You will do this four different times—once for each moon.

Here is an example using the data for Jupiter's moon Amalthea.
Amalthea's orbital radius = 1.8×10^5 so r^3 = 5.8×10^{15} km^3
Amalthea's orbital period = 0.5 days, so P^2 = 0.25 days^2

$$Mass_{Jupiter} = (7.9 \times 10^{10}) * (r)^3/(P)^2$$

$$= (7.9 \times 10^{10} \; kg * day^2/km^3) * \frac{(5.8 \times 10^{15} \; km^3)}{(0.25 \; days^2)}$$

$$= 1.83 \times 10^{27} \; kg$$

With this equation and the values you have measured for the periods of Jupiter's Galilean moons, use the worksheet below to calculate Jupiter's mass four different times. After you have done this, you will calculate the average of your values for Jupiter's mass and find out just how close you came.

Use this worksheet to make your calculations:

Io: $M_{jup}{}^1 = (7.9 \times 10^{10}) * (r^3)/(P^2)$

$= (7.9 \times 10^{10}) * \dfrac{(\underline{\hspace{2cm}})}{(\underline{\hspace{2cm}})}$

$= (\underline{\hspace{2cm}})$ kg

Europa: $M_{jup}{}^2 = (7.9 \times 10^{10}) * (r^3)/(P^2)$

$= (7.9 \times 10^{10}) * \dfrac{(\underline{\hspace{2cm}})}{(\underline{\hspace{2cm}})}$

$= (\underline{\hspace{2cm}})$ kg

Ganymede: $M_{jup}{}^3 = (7.9 \times 10^{10}) * (r^3)/(P^2)$

$= (7.9 \times 10^{10}) * \dfrac{(\underline{\hspace{2cm}})}{(\underline{\hspace{2cm}})}$

$= (\underline{\hspace{2cm}})$ kg

Callisto: $M_{jup}{}^4 = (7.9 \times 10^{10}) * (r^3)/(P^2)$

$= (7.9 \times 10^{10}) * \dfrac{(\underline{\hspace{2cm}})}{(\underline{\hspace{2cm}})}$

$= (\underline{\hspace{2cm}})$ kg

Round the answers to 2 decimal places and record them in the data table.

Average mass of Jupiter: $= \dfrac{M_{jup}{}^1 + M_{jup}{}^2 + M_{jup}{}^3 + M_{jup}{}^4}{4}$

$= \dfrac{(\underline{\hspace{1.5cm}}) + (\underline{\hspace{1.5cm}}) + (\underline{\hspace{1.5cm}}) + (\underline{\hspace{1.5cm}})}{4}$

$= (\underline{\hspace{2cm}})$ kg

Section 10.6

Now you need to calculate your accuracy by finding the percent error for your measurements as follows:

Average % Error $= \dfrac{|(\text{Accepted Value}) - (\text{Average Measured Value})|}{(\text{Accepted Value})} * 100$

The Accepted Value (determined by very accurate measurements made by the Voyager spacecraft when they flew past Jupiter) is 1.90×10^{27} kg. Using the example data given above for Amalthea, we can calculate the percent error for our example.

% Error $= \dfrac{|(\text{Accepted Value}) - (\text{Measured Value})|}{(\text{Accepted Value})} * 100$

% Error $= \dfrac{|(1.90 \times 10^{27}) - (1.84 \times 10^{27})|}{(1.90 \times 10^{27})} * 100$

$= \dfrac{(0.06 \times 10^{27})}{(1.90 \times 10^{27})} * 100$

$= (0.03) * 100$

$= 3 \%$

Now calculate the percent error for your Average Measured Value in the same way.

Average % Error $= \dfrac{|(1.90 \times 10^{27} - ())|}{(1.90 \times 10^{27})} * 100$

$= \dfrac{()}{(1.90 \times 10^{27})} * 100$

$= () * 100$

$= () \%$

Round the answer to the nearest whole number, and record it in the data table.

Data Table

Jupiter's Moons						
Moon	Orbital Period (days)	P^2	Orbital Radius (km)	R^3	Jupiter's Calculated Mass	% Error
Io			4.2×10^5			
Europa			6.7×10^5			
Ganymede			10.7×10^5			
Callisto			18.8×10^5			
Average						

Questions

Q1 What is the average percent error of your measurements?

Q2 Are all your answers close to one another?

Q3 Which measurement was the farthest off?

Q4 Suggest a reason to explain the poor accuracy of this measurement.

Q5 If an object is so far away in space that we cannot send a spacecraft to it, what two things would we have to be able to measure to calculate its mass? (Look back at the equation we used to calculate Jupiter's mass long before we even had spacecraft.)

Q6 Before 1978, we did not know that Pluto had a moon. Before this discovery, could we have known what its mass was?

After we were able to measure the period and orbital radius of Pluto's tiny moon, Charon, we discovered that Pluto was much less massive than we had originally thought.

Q7 As we look even farther than the edge of our solar system, we find that we can determine the mass of many stars. Which of the following stars would have masses that could be measured?
1) Single stars
2) Double stars
3) Individual stars in a star cluster
4) None of the above

Q8 The Andromeda galaxy has two small companion galaxies that orbit it. Could we determine the mass of the Andromeda galaxy knowing the orbital period and distance of these satellite galaxies?

Q9 We can often measure the speed of a star orbiting a star cluster. If we also know the distance at which the star orbits, we can then calculate how long it would take for the star to make one orbit (even though we will not be around long enough to observe one complete orbit). With this information, how would you deduce the mass of the cluster?

Q10 Do you need to know the mass of a moon in order to calculate the mass of the planet it is orbiting?

Section 10.7

To see how astronomers use this equation to measure objects clear across the universe, select *RedShift* 3/*RedShift* **college edition** from the Window menu to reopen the *RedShift College Edition* window. Click on the **Go** button in the navigation bar and select **Principia Mathematica, Cepheid Variables and the Doppler Shift**, and finally click on page **24**. Work through the problem presented there and on page **25**. Then continue through page **29**.

Discussion

With Newton's laws, we can calculate the mass of any object that has a natural satellite orbiting it if we can measure the period of the orbiting object and the radial distance from its parent body. Thus, we can measure the mass of planets (including Earth) using moon motions, and we can measure the mass of stars (including our Sun) using planet motions.

The same rule applies to stars or gas clouds orbiting a galaxy. It even applies to galaxies orbiting the center of a galaxy cluster. In short, we can measure the mass of just about anything if we can find something orbiting it.

Conclusion

On a separate sheet of paper, describe the process of determining the mass of an object in space. List what you must know in order to measure the object's mass. Also, describe the factors that affect the accuracy of your measurements. Finally, suggest some ways you might improve the accuracy of your measurements.

Name: _____

Course: _____

Section: _____

There's No Place Like Home

Introduction and Purpose

In all our explorations of the Solar System and the rest of the universe, it is easy to forget that our own Earth is an important place in space, too. We have learned most of what we know about the universe by observing behaviors, relationships, and theories here on the Earth.

The Earth turns out to be an ideal laboratory for investigating the forces and processes that have shaped our solar system. It has an example or two of almost every kind of feature found on any planet or moon we have yet explored. For this reason, we are going to begin our exploration of the worlds of the Solar System with the Earth. Here, we can get a close-up look at most of the things we will encounter later on as we look at the other planets.

Our purpose in today's exercise is to take an overview of the different physical aspects and characteristics of the Earth as a planet. We will see just what kind of features we find here and in roughly what proportions. Then we will have a foundation for understanding what we see in the rest of the Solar System. Finally, we will take a look at some features *not* found anywhere else that combine to make the Earth the unique and wonderful environment it is.

Procedure

Section 11.1

Start *RedShift College Edition*, and then bring the *RedShift 3* screen to the front. From the Information menu, open the Photo Gallery. In the Photo Gallery Contents frame, open the **Solar System** folder and select **Earth**. When the list of Earth photos appears in the window, scroll down to the image titled **[0699] Earth from space**. Enlarge the photo, and answer the following questions.

Q1   List the colors you see in this photo.

Q2   Tell what kind of terrain or phenomena is related to each color you listed in the previous question.

Q3   We could divide these colors and aspects of the Earth into three main categories: water, land, and atmosphere. Each plays a vital role in forming

11

the environment we live in. List at least one way in which each of these three components makes our world suitable for life.

Q4 Do you see more blue or more brown?

Section 11.2

Click the Thumbnail button to go back to the Earth photo list, and then close the Photo Gallery window by clicking on the close button at the top right corner of the window. Select **Base Plane** from the Control menu, and click on **Eq. Earth**. Click on the large Location button in the Location panel and select **Space: track body**. If the Lon (longitude) and Lat (latitude) boxes are not visible in the Location panel, click on the small icon at the upper right corner of the Location panel to expand it. Now click on the small arrow to the right of the Location box, choose **Planets/Moons**, select **Earth**, and **Lock** it in the center. Set the panel values as shown in the table below, and press the keyboard <Enter> key.

Lat (Latitude)	00° 00′ 00″
Step	1 hour
Zoom	1

Q5 Click on the Forward Step button in the Control Time panel repeatedly until you have gone all the way around the Earth. Is there more water on the surface of Earth or more land?

Q6 Is the land all in one area surrounded by ocean, or is it broken into large sections scattered all over?

Q7 Suppose there were no water on the surface of the Earth. Would you still be able to make out the continents?

Q8 If there were no water on the Earth's surface, what would the continents be like? (How would you describe the continents if you lived in one of the ocean basins?)

Q9 Now click on the Filters button, and select **Sun & Planets**. At the right side of the Filters window, click on the **Phases**, **Atmosphere**, and **Features** boxes to remove those check marks, and then click **OK**. Now change the display parameters as shown in the following table, and press the keyboard <Enter> key.

Date	11/02/1999
Universal Time	14:00
Lon (Longitude)	275° 00′ 00″
Lat (Latitude)	+25° 00′ 00″
Zoom	3.5

Now examine the west coast of North America and look for extended linear features (mountain chains). In which direction are they running?

Q10 Now change the latitude (Lat) to **−25° 00′ 00″** and the longitude (Lon) to **320° 00′ 00″**, and press the <Enter> key. Do you see another linear counterpart for the mountain chains in North America? What is this mountain chain called?

Q11 Suppose that the western coast of the Americas was a car bumper. If you were to see a car bumper with such a long, linear crumple, what would you think had happened to the car?

Q12 Change the Zoom to **1**, and press the <Enter> key. Now, rotate the Earth until you can see both sides of the Atlantic Ocean. Notice the shape of the western coast of Africa and the eastern coast of South America. If the continents were puzzle pieces, what would you deduce about these two?

Section 11.3   Web Intrigues
(If you do not have Internet access, you may skip this question.)
Log on to the Internet, and open your web browser. Type in the following URL: *http://www.fourmilab.to/earthview/* to open the Earth/Moon Viewer.

When the Earth Viewer page has finished loading, click on the first link titled **Map of the Earth**. This will load a map of the entire Earth with the clouds removed, showing the continents and oceans as well as which areas are currently in daylight and which ones are in darkness. This is a really amazing view of the Earth using actual satellite photographs, so spend a little time just looking at the various continents. When you are through cruising around the map, click in the center of the Atlantic Ocean, about halfway between North America and northern Africa. This will zoom the view screen in on that area. If it is currently night in that area, scroll to the bottom of the window, click in the box titled **No night**, and then click the **Update** button to show the area in full sunlight.

11

Q13 Describe what you see curving back and forth down the center of the view screen.

Q14 This image does not show the ocean floor but does attempt to indicate the depth of the ocean with different colors. Based on the coloration of this image, does this curving line look as though it is raised above the ocean floor or sunk into it?

Q15 Are there any places where this line rises above the ocean surface? What would these places be?

Q16 Describe any features that occasionally extend at right angles from this curving line.

Now click on the **Back** button in the browser to return to the whole-Earth map. Scroll down below the map, click in the radio button titled **Topo map**, click in the **No night** box, and then click the **Update** button to drain the water right out of the oceans! (A **Topo map** is a *topographic* map— one that shows elevations as well as shapes.) This time, you will see the solid surface of the Earth without any clouds or water to cover it up. When the map has loaded, again click halfway between North America and northern Africa to zoom in. (Once again, you may need to click the **No night** box and update the view screen.)

Q17 Now look at the curving line through the center of the Atlantic Ocean. Was your guess in Q14 correct?

Q18 Look closely at the section near the center. Does the line appear to be a single ridge or a pair of ridges running parallel to each other?

Q19 Now click on the Back button to go back to the whole Earth Topo map. Look around at the other ocean basins. Which other oceans have similar curving lines?

As you saw, the ocean floor is pretty weird. In many places around the world, you see what look like poorly sewn surgery incisions. One of the most incredible of these scars runs from Iceland in the far northern Atlantic and parallels the coast of Europe and Africa all the way to Antarctica. This long, twisting line is called the Mid-Atlantic ridge. It is essentially just what it looks like—an incision in the crust of the Earth—except that this incision has been there and has been spreading for tens of thousands of years. The west coast of Europe and Africa matches the east coast of the Americas for a very good reason—they were once a part of the same land mass, as has been shown by similar fossil records. Long ago, however, the crust beneath these continents split and forced them apart.

If North America is being pushed westward and Europe is being pushed eastward, something has to give somewhere else. North America has in fact run into something coming from the other direction, the Pacific plate. Where the two met, the collision wrinkled the bumper of the Americas into the immense mountain chains of the Sierras and the Andes.

Section 11.4

Q20 What do you think is causing the crust of the Earth to swell up all along the Mid-Atlantic ridge?

Q21 Do you think the inside of the Earth is hot or cool? Give a reason for your answer.

The volcanoes that pop up here and there are just *small* vents for the enormous amount of pressure that builds up under the crust because of the Earth's hot center. The real pressure-relief valves are these thousands of miles of sub-ocean ridges that allow the pressure to actually push large pieces of the crust around. These large areas of the crust that "float" around on top of the mantle are called *plates*. Volcanoes, earthquakes, continents, and plates all have the same cause—heat rising from the center of the Earth. They are all lumped together in one broad category called *tectonism*. The fact that Earth has continents and expansive, deep ocean basins is a reflection of the role that *tectonic* activity has played in shaping the Earth. Volcanoes, earthquakes, mountain chains, and even the continents rising high above the ocean floor are all the result of heat from the inside of the Earth. While we have not been able to drill into the Earth very far, we have a good idea of what is inside, thanks to earthquake waves that go all the way through the Earth.

A doctor who wants to see how a baby is progressing inside a pregnant woman uses an amazing tool called a sonograph. The sonograph sends vibrations into the woman, and from the echoes that come back, the machine can reconstruct a 3-D picture of the baby inside. In the past 50 years or so, scientists have been using vibrations from earthquakes and man-made detonations to probe the interior of the Earth in the same way. Let's see what they have found.

11

Click the small **x** at the top right corner of the web browser window to close it. *(Macintosh users will click the small square in the upper left corner of the window to close it.)* Now click on the small arrow to the right of the Aim box, choose **Planets/Moons**, and then click on **Earth** to aim the view screen at Earth. You may still see stars on the view screen, but now if you click anywhere on the sky, an information tag titled **Planet: Earth** will pop up. Click on the tag to open Earth's information box, and then click on the **More** button. When the information box expands, click on the tab titled **Physical Data** and study the diagram at the right of the box showing the interior layers of the Earth.

Q22 Name the four layers of the Earth in order, beginning from the surface.

Q23 Based on our understanding of how shock waves move through the Earth, we have concluded that each of these layers is made of a different material. Think about your experience with stuff that floats. If the crust is floating on top of the mantle, is the material of the crust lighter or heavier than the material of the mantle?

Q24 If all the layers beneath the crust are semi-fluid, where would the heaviest material in the Earth end up?

Q25 What characteristic of the inside of the Earth could cause rock and metal to become fluid or semi-fluid?

Because of the tremendous amount of heat coming up from the center of the Earth, the surface of our planet has been riddled by volcanoes, split by earthquakes, and wrinkled into mountain chains by colliding continents. In the next exercise, we will be looking at various features of planet surfaces. We will find there are three major shapers of surfaces: impacts, tectonic activity, and flow features. The fact that Earth has continents and expansive, deep ocean basins is a reflection of the role that tectonic activity has played in shaping our planet.

Section 11.5

Now let's explore some of the other characteristics of the Earth's surface. Reopen the Photo Gallery from the Information menu. Again, choose **Solar System** and select **Earth from space**. This time, scroll down to the image titled **[1436] Gora Konder crater in Russia**, enlarge the photo, and answer the following questions.

Q26 In this photo, what formed the extensive forking patterns?

Q27 In which direction does the major streambed in the lower right-center of the picture lead?

Q28 Look closely and take a guess as to whether these streams originated in springs or from rainfall runoff.

Q29 Click on the Thumbnail button, scroll down to the image titled **[1407] Mauna Loa volcano on Hawaii**, and enlarge it. Describe any features in this photo similar to the ones you examined in the previous photo.

Q30 Describe any major differences between these erosion features and those in the previous picture.

Q31 Click on the Thumbnail button, and read the caption accompanying this photo. Are these erosion features formed by water, as were those in the previous photo?

Q32 Scroll on down to the image titled **[0702] Nile River Delta**, and enlarge it. This is the mouth of a very large river. What is happening to the features originally at the bottom of the ocean directly in front of this outflow?

Q33 The water on Earth plays another very critical role besides that of supporting life as we know it. The presence of so much water on the surface of the Earth (along with water vapor in the atmosphere) plays a huge role in moderating Earth's temperature.

To understand how this works, remember the last time you were at a lake on a hot summer day. In the afternoon of a long, hot sunny day, is the temperature of the dry sand hot or cool?

Q34 Is the temperature of the water hot or cool?

Q35 If you go back to the lakeshore early the next morning, however, things are different. What is the temperature of the sand then?

Q36 And what about the water? Did its temperature change as much as that of the sand?

11

Q37 These answers illustrate on a small scale the effect of heat on water. A long, hot day does little to warm the lake, but it heats the sandy earth to blazing temperatures. The night cools the sand just as quickly as the Sun warmed it, but again it does little to change the temperature of the water, which then (in contrast to the cool air and sand) seems warm the next morning. Given your answer to Q5 about the amount of water on the surface of the Earth, what do you think the temperatures on Earth would be like if we had no oceans?

Section 11.6

As you have seen, the surface of the Earth is also shaped extensively (though on a much more local scale) by the movement of fluids across its surface—the process of erosion. On the Earth, most erosion is done by water, but in some regions and on several other bodies of the Solar System, the erosion is carried out by lava. No matter how it is done, the end result is a wearing down of some features (like the impact crater in the first photo) and the covering up of others (like the ocean bottom under river deltas). If the flow is extensive and long-lived, erosion and resurfacing can nearly obliterate the original surface of a planet or moon.

Click on the Thumbnail button, and scroll back up to the image titled **[0709] Great cyclonic storm**. Enlarge this picture, and answer the following questions.

Q38 In this picture, you are looking at the top of clouds in the upper atmosphere of Earth. What do we call the feature that is most prominent in this photo?

Q39 Cyclonic storms on Earth are caused by a combination of high and low pressure areas (caused by unequal surface heating) and the rotation of the Earth. The combination of the rotation of the Earth and a second, perpendicular motion is known as the Coriolis Force (after the person who first explained it). In the case of the atmosphere, this phenomenon usually occurs as cool air moves perpendicular to the direction of the Earth's rotation. In the Northern Hemisphere, would the cool air be coming from the north or the south?

Q40 As this cool air moves down across the Earth's surface, the surface is rotating underneath it. In which direction is the surface of the Earth moving—to the east or to the west? (Think about how the Sun comes up.)

Q41   Now, in your mind's eye, imagine the air moving from north to south with the Earth rotating from west to east underneath it. Which way will the air flow curve, to the right or left?

Q42   As the air moves further south, it is heated and begins to rise and move back northward. The result is a circular motion. Will this circular flow be clockwise or counterclockwise in the Northern Hemisphere?

Q43   The Earth turns the same direction in the southern hemisphere as in the northern. So why do cyclones in the Southern Hemisphere turn in the opposite direction?

To see how this works, click the **Go** button in the navigation bar, select **Science of Astronomy**, **Worlds of Space**, **Worlds in the Clouds**, and click on page **16**. Now, read and watch through page **18**.

Q44   Click the **Science of Astronomy** selection tab in the navigation bar, and choose **Photos** again. In the Contents box, select **Earth** in the **Solar System** section, and click on **Earth from space**. Scroll down to the image titled **[0703] Southern California, Baja, from Apollo 16**. Enlarge the image. How many areas of cyclonic circulation can you see in this photo?

Q45   As you have scrolled through these images of Earth from space and as you look at this photo, think about the amount of cloud cover you have seen. Is it large or small?

Q46   Based on your answer to the last question, suggest some ways you think clouds might affect overall conditions on Earth.

Q47   Besides giving us something to breathe and supporting clouds, our atmosphere protects us as well. From the Window menu, select *RedShift 3/RedShift* **college edition** to reopen the *RedShift College Edition* window. Click on the **Go** button in the Navigation bar, select **Science of Astronomy**, **Worlds of Space**, and **Moons**, and click on page **2**. Read, watch, and do the simulations through page **9** to learn about some of the protective aspects of our atmosphere. Combine this information with what you've learned about the surface-shaping forces we have discussed, and list two reasons why we don't see many meteor craters on the Earth today.

11

Discussion

Our Earth is a showcase for three major components of most planetary bodies—solid, rocky surfaces; water (and its accompanying effects); and atmospheres. The surface geology of Earth gives us a wonderful jumping-off point for exploring the surfaces of the other terrestrial planets and most of the moons in our solar system. In the next exercise, we will see that we find nearly all of the features seen on Earth in other places as well. As we try to isolate the causes of these features in the exercise "Fire and Brimstone," we will take a significant step toward understanding how our solar system came to be as it is.

We used to think that water was a rare commodity and that Earth was a unique place to have so much of it. Now we know otherwise. Water is one of the most common compounds in our solar system. Most of the water in other places, however, is not in liquid form. It exists in most places either as water vapor or as ice. The effects of water on our planet cannot be overstressed.

First—and most obvious—water is vital to life as we know it. Our bodies are eighty percent water, and without it we would die in just a few days. Second, water moderates Earth's temperatures. Suppose the whole Earth had temperature extremes like the great deserts, with temperatures well over 100 degrees Fahrenheit in the daytime and below freezing at night. And finally, water tailors our atmosphere.

Then we come to our atmosphere. Our atmosphere not only gives us something to breathe, but it aids us in many other ways as well. It supports clouds that circulate our water from ocean to continent. Through the water vapor in the clouds, it helps moderate Earth's temperatures. Finally, it protects us from harmful X-rays and ultraviolet radiation from the Sun as well as from the tons of meteoric debris that fall on the Earth every day.

Conclusion

On a separate sheet of paper, summarize what you have learned today about the overall characteristics of Earth. Describe its surface makeup, listing various common features. Describe the effects of tectonic activity such as volcanism and continental drift and of erosion on the surface features of Earth. Finally, describe the role that the oceans and atmosphere play in creating an environment perfectly suited for life.

Name: _____

Course: _____

Section: _____

"Hardbodies"

Introduction and Purpose

The planets of our solar system are divided into two distinct groups—the terrestrial planets (or inner planets) and the Jovian planets (or outer planets). The makeup of these groups is very different. Studying the differences and similarities among the planets in these two groups tells us a great deal about how our solar system formed and how the planets came to be as they are. Today, we will examine the "hardbodies" of the Solar System, the terrestrial planets.

The word *terrestrial* comes from the Latin name for Earth, "Terra." The terrestrial planets are Mercury, Venus, Earth and Mars. In today's exercise, we are also going to include the Moon in our discussion. The hard surfaces of these bodies have been shaped mainly by three mechanisms.

First, they are shaped by impacts from meteors, comets, and asteroids and their fallout. Second, the hard surfaces of a planet or moon can also be shaped by tectonic forces from within. Internal heat can cause the solid surfaces of these bodies to split, fracture, swell, warp, or pop open in a planetary pimple—a volcano. Finally, the surfaces of some of these bodies are sculpted by fluid flow. Both atmospheric flow (wind) and surface liquid flow (lava or water) can smooth edges, erode older features, and cut sinuous valleys and canyons across craters, mountains, and smooth plains.

In today's exercise, we will look at the physical characteristics of the surfaces of the terrestrial planets and see what similarities and differences we can find. We will categorize the features we find as impact features, tectonic features, or flow features, and see which features are most common on which planets.

Procedure

Section 12.1

Start *RedShift College Edition*. If the Contents frame is not open, click on the **Contents** button in the navigation bar to open it. Now click on the **Science of Astronomy** selection tab, and choose **Photos** to open the Photo Gallery window. In the contents frame, open the **Solar System** folder and select **Mercury**. Then answer the following questions.

Q1 In the Mercury photo window, find the image titled **[0877] Mercury's southern hemisphere (Mariner 10)**. Click on the Magnify button to

12

enlarge the image, and study it for a moment. What is your first impression?

Q2 In this image of Mercury, there are many straight lines emanating from craters. These are called "rays." In the previous exercise, you looked at a cratering simulation from one of the chapters in the Science of Astronomy. There you saw that these rays were formed by bright subsurface material being "splashed" over the darker surface. What does this information suggest about the age of craters with the streaks emanating from them?

Q3 Look carefully at the photograph again. Do you see any linear features that do *not* seem to be coming from any craters? (These would be fault lines or mountain chains.)

Q4 Click on the Thumbnail icon to go back to the Mercury window. Scroll down to the photo titled **[0028] The southwestern quadrant of Mercury**. Enlarge this photo and study it carefully, looking for noncrater linear features. In this photo, you should be able to find at least one linear feature in the lower left center. (There are others as well.) This feature is evidence of stresses from inside the planet. These kinds of features are called *tectonic* features. Write a phrase or two in the second column of the data table summarizing any such features you can find.

Q5 Click on the Thumbnail icon to go back to the Mercury window. Scroll down to the photo titled **[2019] Part of the Caloris basin on Mercury**. Enlarge this photo, and study it carefully. In this photo, you should see a huge semi-circular feature at the left of the image. This is the edge of the Caloris Basin. In this basin are several concentric rings. How many distinct rings can you see?

Q6 Look at the ordinary craters in this photo. Is the Caloris basin younger than the rest of the craters, or older?

Q7 What reason can you give for your answer to the last question?

Q8 On the basis of its shape and size, do you think the Caloris basin was formed by an impact from a huge planetoid or from a volcano? Why or why not?

Summarize the amount and types of impact features seen on Mercury in the first column of the data table.

Q9 In any of these pictures, have you seen features that look as though they were caused by flowing air, water, or lava? If you have, summarize those features in the third column of the data table.

Data Table

Terrestrial Planets (and Moons)			
Planet	Impact Features	Tectonic Features	Flow Features
Mercury			
Venus			
The Moon			
Earth			
Mars			

Section 12.2

In the Contents frame, click on **Venus** to open the Venus photo list and then do the following.

Q10 In the Venus photo window, scroll down to the photo titled **[1197] Gula Mons, Sif Mons, and lava plains on Venus**. As you look at this false-color image, jot down each type of feature you see that might have been caused by tectonic activity. (Look for things like mountain ridges, volcanoes, and fault line fractures.) Summarize these features in a couple of phrases, and write them in the Tectonic Features column of the data table for Venus.

Q11 Describe any flow features you see in this image.

Q12 Click on the Thumbnail icon again. This time, scroll back up to the image titled **[1192] A Venusian lava flow southeast of Navka Planitia**. Do the sides of this dark lava canyon look gentle and sloping or sharp and steep?

Q13 Do you think any other features in this photo might be flow features?

12

What reason can you give for calling them flow features instead of fractures caused by swelling of the surface?

Q14 Click on the Thumbnail icon to go back to the Venus photo list and scroll down to the image titled **[1193] Three meteorite impact craters in the Lavinia region**. Enlarge the image, and look carefully at these three craters. For a close-up view of one of them, click the Thumbnail icon and scroll about halfway back up to the picture **[0722] The Venusian impact crater Aurelia** and enlarge it.

After studying these two images, write down any similarities and differences you find between these and the craters on Mercury and the Moon. Look for things like central peaks, ejecta blankets and rays, sharp or smooth edges, etc. Go back to either image to refresh your memory if necessary.

Similarities	Differences

Summarize the flow and impact features of Venus in the appropriate columns of the data table in Venus's line.

Q15 Now click on the Thumbnail icon and scroll back up to the image titled **[0203] Dome-like hills on Venus**. Describe these hills, and suggest a theory for the way they might have formed.

Q16 Describe any evidence of fracturing on the tops of these hills or in the surrounding area.

Q17 Click on the Thumbnail icon again. This time, scroll all the way to the top to the image titled **[1198] Venus from Magellan radar images**. Study this image carefully, looking for all the things we have been talking about—mountain ranges, fracture lines, volcanoes, and meteor impact craters. Do you see more evidence of impact activity (meteor craters), tectonic activity (volcanoes, faults, and mountain ridges) or flow activity?

Q18 Based on your observations in the preceding question, propose some differences between the interior of Venus and that of Mercury.

To see just how amazing the terrain of Venus is, click the **Go** button in the navigation bar, select **Science of Astronomy**, **Worlds of Space**, **Hard Shells/Soft Centers**, and click on page **9**. Now, watch this movie of a fly-over of a part of the surface of Venus. (You can pause the movie at any point with the **Pause** button or replay the movie by clicking on the **Replay** button in the navigation bar.)

Section 12.3

Click on the **Science of Astronomy** selection tab, and again select **Photos**. In the Contents frame, select **Moon** from the Solar System section.

Q19 Scroll through the Moon photo list, and click on the one titled **[1189] A gibbous Moon**. Look carefully at this picture, and then compare it with the pictures of Venus and Mercury. In the space below, write down any similarities you can find with the photos of Venus or Mercury, and then note any differences.

Similarities   Differences

Q20 Look carefully to see whether you can discover any kind of feature you did *not* see on either Mercury or Venus. (Do you see anything that is not a crater, mountain ridge, volcano, or canyon?) List any features that fit this description.

Q21 Click on the Thumbnail icon and scroll down to the photo titled **[1520] Apollo 11 lunar surface scene.** In this photo, there are lots of craters. Do they look more like the craters on Mercury or on Venus?

Q22 Some of these craters are very old, and some are much younger. The younger ones are on top of the older ones, obviously. Can you find any distinguishing features that might help to determine the age of a crater? (Look at the texture and smoothness of the edges of the craters, and compare those on top with those underneath.)

Q23 The very large crater at the top of the photo is called Daedalus. Large craters are usually made by the impact of really big objects. The enormous release of energy accompanying these collisions not only makes a bigger crater, but it also imparts to the crater certain characteristics that are different from those of smaller craters.

141

12

In this picture, compare Crater Daedalus with the smaller craters. Note the smoothness or roughness of the crater walls, the presence or absence of concentric rings or terraces within the crater, and the presence or absence of a central peak.

Daedalus   Other Craters

Q24   Click on the Thumbnail icon and scroll down to the image titled **[1513] Apollo 17 extravehicular activity photomosaic**. Enlarge this photo, and look at the surface of the Moon near the Apollo 17 landing site. List any features that might have been caused by impact and any that might have been caused by tectonic activity.

Impact Features   Tectonic Features

Q25   Click on the Thumbnail icon, scroll down to the image titled **[0304] The lunar surface**, and enlarge it. List any features that you can find that are *not* related to impact craters.

In the appropriate columns of the data table, summarize the Moon's impact, tectonic, and flow features.

Section 12.4

In the **Contents** frame, choose **Earth**, and then click on **Earth from space** to open a new photo list. Answer the following questions.

Q26   Scroll down the image list to the photo titled **[1457] Southern Washington state**. Enlarge this photo, and look at the surface of the Earth. In this picture, you can see three volcanoes amid typical terrestrial geography. Compare this view with the image you looked at previously of the volcanoes and their surroundings on Venus. What features do you find in both? (Go back and look at the picture **[1197] Gula Mons, Sif Mons and lava plains on Venus** if you don't remember it well enough to compare.)

Q27   Scroll down a little farther to the photo titled **[1427] Tambora Caldera, Sumbawa Island, Indonesia**. Enlarge this photo, and compare this volcanic crater (caldera) with the impact craters you have seen on other bodies of the Solar System. List any features you can see that might allow a person to distinguish between a caldera and an impact crater.

Q28 Scroll back up to the photo titled **[1436] Gora Konder crater in Russia**. Enlarge this photo, and compare it with the previous one and with the photos of impact craters on the other planets. This is an impact crater caused by a meteorite striking the Earth. List any similarities you see between this crater and the impact craters you have observed on the Moon and the other planets.

Q29 How is this crater different from the caldera in Indonesia?

Q30 There is a flow feature running out from this crater. How do you think this valley might have formed?

On the basis of these photos and your own observations of Earth, summarize the commonness of each kind of feature (impact, tectonic and flow) on the Earth in the appropriate columns of the data table.

Section 12.5

Click on the Thumbnail icon to go back to the list of Earth photos. Now, in the Contents frame, choose **Mars** and click on it. Then answer the following questions.

Q31 Scroll about halfway down the list to the image titled **[0910] Martian impact craters**, and enlarge it. Compare these craters on Mars with those on the other planets. List the main differences you see between these craters and those on the other bodies we've looked at today.

Q32 Click on the Thumbnail icon, and scroll down to the image titled **[0886] Mangala Valles**. Enlarge this image and study it. Does something seem odd about this picture? It looks like stream valleys, doesn't it? But do they seem to protrude from the surface rather than being sunk in? If they do, just step back from the monitor and rotate your head 180 degrees to turn the picture upside down. Now how do they look?

Our brains are funny things. Most of us expect the light to be coming from the top of a picture rather than from underneath, so our brains automatically reverse the image when the light is coming from the wrong way. What would you say caused these sinuous valleys?

Q33 Are these valleys older or younger than most of the craters in this picture?

12

How can you tell?

Q34 Click on the Thumbnail icon, scroll down to the image titled **[0887] Conical volcanoes and fossae on the Tharsis Ridge** and enlarge it. In this picture are several volcanoes with caldera craters. Compare these with the caldera in Indonesia on Earth, again listing the main features of a caldera.

Summarize the three types of features as found on Mars in the appropriate spaces in the data table.

Section 12.6

Q35 Click on the Thumbnail icon, scroll down to the image titled **[3028] Enhanced mosaic of Viking images of Mars**, and enlarge it. Look for evidence of the three types of surface features we have been finding.

Notice especially the long, straight, horizontal trench in the lower center of the photo. This is Valles Marineris. It looks a lot like the top of a loaf of bread that split as the dough inside rose faster than the outside could expand. This is exactly what we believe caused this enormous canyon. Internal heated matter rose near the surface on this side of Mars splitting the surface along Valles Marineris and erupting as volcanoes in the region just to the left. List as many features as you can find in this photo for each category.

Impact features   Tectonic features   Flow features

Q36 Click on the Thumbnail icon, scroll down to the image titled **[3029] Enhanced view of Mars from Viking images**, and enlarge it. Study this picture, looking for evidence of the three types of surface features we have been finding. This is the other side of the planet from the picture we just looked at. What kinds of features are most prominent in this hemisphere—impact, tectonic, or flow?

Q37 Click on the Thumbnail icon to return to the photo list. Was there a big difference in the distribution of types of features on the two different hemispheres of Mars?

Is it reasonable to think that perhaps all the meteorites hit Mars on the same side and left the other side untouched?

Q38   If it is not reasonable to think that one side of Mars would be virtually untouched by meteorites, propose an explanation for the scarcity of impact craters on the hemisphere with the volcanoes, canyons, and flow features.

To get a better idea of the immensity of Mars's canyons and volcanoes, click the **Go** button in the navigation bar, select **Science of Astronomy**, **Worlds of Space**, **Hard Shells/Soft Centers** and click on page **25**. Now watch this movie of a flyover of a part of the surface of Mars. (You can pause the movie at any point with the **Pause** button or replay the movie by clicking on the **Replay** button in the navigation bar.)

Web Intrigues

(If you do not have Internet access, you may skip this question.)
Log on to the Internet and open your web browser. Now, let's see what we are currently learning about our neighbors. Type in the following URL: *http://www.planetscapes.com/solar/eng/homepage.htm* to open the "Views of the Solar System" web site. Once the main page has opened, select either **Mars** or **Venus**.

Venus: If you selected Venus, scroll past the data sheet to the photos, and select several to study and to read the captions. If you want to read more or download a larger version of the photo, click on the photo itself.

Q39   Look through this page to find the answer to this question: What is an ***arachnoid***?

When you have thoroughly explored this site, type in the following URL: *http://nssdc.gsfc.nasa.gov/photo_gallery/* to open the NSSDC photo gallery page. Now choose **Venus** to discover more about the surface features of Venus. Scroll past the photos, and click on the link titled "About the Magellan mission" to find the type of radar used by the Magellan spacecraft to take all the 3-D images. Write the type of radar here.

Mars: If you selected Mars, scroll past the data sheet to the photos and select several to study and to read the captions. If you want to read more or download a large version of the photo, click on the photo itself.

12

When you have thoroughly explored this site, scroll back to the top of the page. Look through the table of contents box at the left of the page, find the **Mars Resources** section, and click on the link titled **Mars Pathfinder Mission**. Once the Mars Pathfinder page has opened, click on the link titled **Mars Pathfinder Web Site at the End of the Mission**. Now explore the fabulous surface photos taken by the Pathfinder. (Be sure to click on the photos to enlarge them.) Also check out the **Archive of all Images** link on this page. If you have a Virtual Reality Viewer in your browser, look at the **VRML** page as well. There is also a large selection of very cool 3-D images here if you have some red and green 3-D glasses.

Q39 What is the name of the two prominent mountains visible from the Mars Pathfinder landing site?

Q40 What was the name of the Mars Rover?

When you have read through these web sites, write a paragraph describing our current research and experiments on the planet you chose.

Section 12.7

For a good review of the terrestrial planets our solar system, reopen the *RedShift College Edition* window, click on the selection tab, and select **Science of Astronomy** from the drop-down menu. Then, in the Contents frame, select and go through the segment titled **Hard Shells/Soft Centers** in the **Worlds of Space** chapter. As you go through these pages, look for evidence of the three planet-shaping processes: impacts, tectonic activity, and fluid flow (erosion and deposition).

Discussion

In today's exercise, we compared and contrasted the surfaces of the terrestrial planets and of our own Moon. We found evidence of all three surface shapers—impacts, tectonic activity, and fluid flow—on nearly every body we studied. On some, however, impacts were fairly scarce; on others, flow features and tectonic activity were difficult to find. We will analyze some of the other characteristics of these bodies in the next exercise to see if we can find a reason for the different distributions of these features.

Conclusion

On the next page, write a paragraph summarizing each of the three mechanisms for shaping the surface of a planet or moon. Describe the visible effects of each mechanism. Give a couple of specific examples of each type as observed in today's exercise.

Describe how you would distinguish between the tectonic effects of caldera craters and rift valleys and their nontectonic look-alikes—impact craters and river valleys. Where several actions have taken place in the same vicinity on a planet's surface, describe how you might decide which was older and which more recent. (Continue your conclusion on a separate sheet of paper if necessary.)

Finally, fill in the following chart ranking the terrestrial planets and the Moon. In the first column, rank them in order of the commonness of impact features on each body, listing the body with the most impact features first and then going on down to the one with the least impact features. Then do the same for tectonic features, ranking the bodies from the one with the most to the one (or ones) with the least. Finally, rank each body according to the prominence of its flow features.

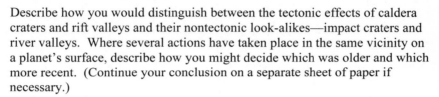

Terrestrial Bodies Ranked by Surface Features		
Impact Features	Tectonic Features	Flow Features
1)	1)	1)
2)	2)	2)
3)	3)	3)
4)	4)	4)
5)	5)	5)

12

Notes

Name: _____

Course: _____

Section: _____

Fire and Brimstone

Introduction and Purpose

One of the most violent and spectacular natural occurrences on Earth is the eruption of a volcano. Volcanoes, however, are more than just a source of fireworks here on Earth. They turn up on other planets and even on one moon. The cause of volcanism (and of other related tectonic activity, like earthquakes and faults) is of prime interest to modern planetary scientists and is the subject of today's exercise.

As we saw in the exercise on the terrestrial planets, tectonic activity varies from planet to planet. In addition to the planets we examined in that exercise, volcanism is also found on Io, one of Jupiter's moons. The purpose of today's exercise is to try to find what the planets and moons with tectonic activity have in common. Then, perhaps, we can begin to understand why some planets and moons have volcanism and a lot of tectonic activity and others do not.

Procedure
Section 13.1

Start *RedShift College Edition*. Before we go too far into our study, let's review the features that indicate tectonic activity. These are fault lines (and earthquakes), rift valleys, mountain ranges, volcanoes, and continental plateaus. All of these surface features indicate the presence of a fluid or semi-fluid interior at some point in the planet or moon's history. Some, such as fault lines, do not require much fluidity to produce them. Others, like volcanoes and continental plateaus, require high amounts of fluidity.

Web Intrigues

(If you do not have Internet access, you may skip this question.)

Let's begin our overview of tectonics by looking at one of the most obvious indicators of a fluid or semi-fluid interior, the volcano. Open your web browser and log on to the Internet. Now open the ***Volcano World*** URL: *http://volcano.und.nodak.edu/vwdocs/planet_volcano/other_worlds.html* Follow the **Venus** link, and then click on **Large Shield Volcanoes** to learn about Venus's most common volcanoes. After reading about these volcanoes, click the **Back** button on your browser until you return to the **Volcanoes of Other Worlds** page. From this page, follow the links to

13

explore the volcanoes of **Mars** and **Io**.

Section 13.2

When you have finished exploring the volcanoes of other worlds, close your web browser and return to the *RedShift College Edition* screen. Now bring the *RedShift 3* screen to the front. From the Information menu, select **Find** to open the Find window. Click on the small arrow to the right of the **Find:** box, and select **Planets**. In the Planets list, find each planet shown in the table below (**Mercury**, **Venus**, **Earth**, and **Mars**), and double-click on it. When you double-click on the planet name, a list of links will appear in the List Links box. In the List Links box, click on the item titled **Information on...** and then click **OK** to open the information box for the selected planet.

Once the information box is opened, click on the button titled **More** to fully open the information box. In the expanded box, click on the tab titled **Physical Data**. From the information shown there, find the data requested in the data table below, and enter it in the appropriate columns. When you have recorded all the data, click the **Close** button in the information box. Repeat the process for the other planets.

Data Table

Terrestrial Planets and Moons					
Planet	Mass (kg)	Diameter (km)	Density (g/cm^3)	Distance from Sun (AU)	Tectonic Activity
Earth					Drifting continental plates, active volcanoes, mountains, and faults
Venus					Active volcanoes, mountains, and fault fractures
Mars					Dead volcanoes, rift valleys, and faults
Io					Active volcanoes
The Moon					Mountains and small faults
Mercury					Small fault fractures
Ganymede					No significant tectonic activity

Section 13.3

When you have recorded the data for all four of the planets, click on the small arrow to the right of the **Find:** box, and select **Moons** to put the names of the moons in the Find list. In this list, you'll find the remaining objects—the **Moon**, **Io**, and **Ganymede**. Just as before, double-click on the name of the object you wish to learn about, click **Information on...** and then click **OK** to open the object's information box. Again, click on the **More** button and then the **Physical Data** tab to find the information you need to fill in the data table. (Use the Sun's distance to Jupiter for the distance of Io and Ganymede from the Sun. Use the Sun's distance to Earth for the distance of the Moon from the Sun.)

Section 13.4

The data table is arranged in order of the amounts of tectonic activity on each body. The body with the most tectonic activity, Earth, is listed first, and the one with the least amount, Ganymede, is listed last. (Ganymede is included in this list only for purposes of comparison.) The purpose of the first set of questions is to try to find one or more physical characteristics that will help us predict how much tectonic activity a body might be expected to have. Study this table, and answer the following questions.

Q1   In the data table on the previous page, the terrestrial bodies are listed in order of tectonic activity. Look at the other four columns of data. Which other columns are arranged (mostly) in a similar order from largest to smallest?

Q2   Arrange these bodies in order of their densities, from largest to smallest.
1)   2)   3)   4)
5)   6)   7)

Q3   How many of these bodies are not in the same order as in the data table? Name the ones that are out of order?

Q4   Which two of these bodies are most nearly identical in mass and density?

Q5   Are they also nearly identical in their level of tectonic activity?

Q6   Arrange the terrestrial bodies in order of their masses, from largest to smallest.
1)   2)   3)   4)
5)   6)   7)

13

Q7 How many of these bodies are not in the same order as in the data table? Name the one that are out of order.

Q8 Arrange the terrestrial bodies in order of their diameters, from largest to smallest.
1)   2)   3)   4)
5)   6)   7)

Q9 Does the diameter of a body seem to be a consistent predictor of tectonic activity?

Q10 Arrange the terrestrial bodies in order of their distance from the Sun, from closest to farthest away.
1)   2)   3)   4)
5)   6)   7)

Q11 Does the body's distance from the Sun seem to be a consistent predictor of the level of tectonic activity?

Q12 Of these four characteristics—mass, diameter, density, and distance from the Sun—which seems to be the most consistent predictor of the amount of tectonic activity? (Which list is most like the order of the data table?)

Section 13.5

Density is an important characteristic. To help you better understand the concept of density, answer the following children's riddle: "Which is heavier, a pound of lead or a pound of feathers?" (We should write the answer upside-down at the bottom of the page, but we won't!) The answer is, of course, that a pound is a pound is a pound. A pound of feathers is a certain weight, and a pound of lead is the same weight!

The thing about this question that tricks many children (and some adults) is that they know that feathers and lead have very different characteristics. They automatically jump to the wrong conclusion that a pound of feathers (which they know are light) *must* be lighter than a pound of lead (which they know is heavy).

The characteristic that distinguishes lead from feathers, however, is *not* weight; it is ***density***! The correct question would have been, "Which is more ***dense***, a lump of lead or a ball of feathers?" Density is defined as the mass of an object divided by its volume: $D = m/V$. All of the densities you wrote down in the data table were calculated by dividing the masses of the bodies by their volumes.

When you were a kid, did you ever sneak down to the Christmas tree when no one was looking and lift and shake your presents? What were you doing? You were trying to guess what was inside by judging the weight of each present and thus its density. A small, heavy gift meant something neat! A larger, lighter one was probably underwear. By calculating the densities of Solar System objects, we are doing much the same thing. We may never get to "open" some of the bodies we can observe, but we can tell a lot about what is inside them by finding their densities.

To understand just what kinds of substances the planets may be made of, let's look at the densities of a few common Solar System materials.

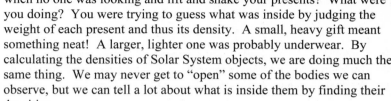

Densities of Selected Solar System Materials (g/cm³)					
Ice	Water	Rock	Iron	Lead	Radioactive Elements (Uranium)
0.9	1	2.5	7.9	11.3	19

Q13 The surface of the Earth is mostly water, with rocky continents coming in second. Of course, the oceans are not very deep (compared with the radius of the Earth), and below them is more rock. Supposing that the Earth were roughly one-tenth water and nine-tenths rock, what should its average density be?

$$\text{Average density} = \frac{(1 * 1) + (9 * 2.5) \text{ g/cm}^3}{10}$$
$$= (\qquad) \text{ g/cm}^3$$

Round your answer off to 1 decimal place.

Q14 Compare your answer to Q13 with the value you recorded in the data table for the density of the Earth. What does this tell you about what is inside the Earth, under the oceans and the continents? Is the interior of the Earth less dense than rock, the same density as rock, or more dense than rock?

Q15 Look at the other materials listed in the table of densities above. Which of them is closest to the overall density of the Earth?

Q16 Another hint about what is in the interior of the Earth comes from looking at the Earth's magnetic field. A magnetic field like that found around the Earth is created by moving electric charge—electrons or protons either

13

spinning or traveling in a circle. The most likely explanation for such a field around the Earth is if the interior of the Earth were made of a conducting fluid, carrying electrons around and around as the Earth turns.

So whatever is inside the Earth must be very dense, must be a good conductor, must be a relatively common material, and must be a liquid at the temperatures found deep below the Earth's surface.

Other than the element mercury, most every common material denser than rock is a solid at ordinary temperatures. If the interior of the Earth must be denser than rock and must also be a fluid, it must be something that is melted. This would require a pretty intense source of heat. List all the evidence you can think of that indicates the interior of the Earth is hot.

Q17 Putting all these criteria together, we find one candidate for the interior of the Earth standing head and shoulders above the rest—iron. Iron is a good conductor, is easily liquefied at moderate temperatures, is one of the most common metals on the Earth's surface, and has a density just a little higher than the density of Earth.

If the Earth were approximately 1/20 water, 8/20 rock, and the rest iron, the overall density would come out just about right. Of course, these percentages are little more than conjecture. Within our Earth, there are lots of different kinds of rock with different densities, as well as many other heavier elements than iron all. The actual makeup of the Earth is therefore impossible to calculate exactly.

This brings us to the most important question of all: Since the Earth is apparently hot inside, where does the heat come from? Can you think of any chemical reaction involving iron that produces heat?

Q18 If the iron does not produce the heat inside the Earth, we must look farther. There is one very simple potential source of heat for the interior of the Earth. As the Earth formed long ago, it was bombarded by meteors, asteroids and even small planetoids. The Earth stopped their motion (and absorbed their energy) just as applying the brakes stops a car. What happens to your brakes when you keep them on a long time (besides wearing out)?

The heavier an object is and the faster it is moving, the more energy it has. When it collides with something else and comes to a halt, that energy has to go somewhere, and most of it usually ends up as heat. Thus every piece of rock that has ever struck the Earth from the very beginning of time has contributed to Earth's residual heat.

Q19 If most of the heat inside the Earth came from those original collisions long ago, what would you think about the temperature inside the Earth now as compared to way back then?

Q20 The energy in a moving object depends not only on its speed, but also on its mass. If most of the heat that causes volcanism and tectonic activity came from the original collisions that formed the planets, describe how this helps explain why more dense planets have more tectonic activity.

Q21 Given your arguments in the last two questions, which would you expect to be hotter, a heavier planet or a lighter one? (Hint: Which one would have absorbed more energy?)

Q22 Which part of a planet would most likely remain the hottest, the surface or the core?

Q23 After many years of cooling, which planet should have a thicker crust, a heavier planet or a lighter one?

Q24 Given your answer to the last question, which planet would you expect to have the most active tectonics today, a heavier planet or a lighter one? (Hint: Would active volcanoes and drifting continental plates be more likely with a thick crust or a thin one?)

Q25 Describe how the preceding arguments help explain why Mercury and the Moon show little tectonic activity.

Q26 Describe how these arguments help explain the differences between Earth and Venus.

Q27 There are other potential sources of heat for the inside of a planet or moon. If the interior of the Earth is mostly liquid iron and you stirred in

13

something like uranium, what would happen to it? (Look at the density of uranium on the chart given earlier.)

Q28   Do radioactive materials produce heat? (Think about the atomic bomb.)

Q29   Which planets would you expect to have more radioactive elements and thus more internal heat, more dense ones or less dense ones?

Q30   Does your answer to the last question agree with what you have observed in today's exercise?

Q31   These arguments related to original heat (and secondary radioactive heating) explain satisfactorily the current state of almost all of the objects we studied today. But where does this argument put Io? If we were to judge Io strictly on the basis of its density, mass, and size, what planet should it be most like today?

Q32   Is Io like this planet?

Section 13.6

To understand why Io is much more active than we think it ought to be, we must look at its environment. While it is likely that Io has an iron or iron sulfide core not unlike the Earth's, it should have lost most of its original heat long ago, and the moon should have solidified. The reason it has not solidified has to do with its orbit around Jupiter.

Jupiter has an even better magnetic field generator than liquid iron spinning around its center. Its insides are mostly liquid and metallic hydrogen, which are some of the best conductors around. As a result, Jupiter's magnetic field is immense. It is still quite strong as it crosses one part of Io's orbit. As a result, Io moves in and out of this magnetic field every 1.7 days. Moving in and out of Jupiter's magnetic field alternately squeezes and then stretches Io like a rubber band.

To see the effect of this, alternately stretch a rubber band and then relax it. After a few times, you should notice a temperature change in the rubber. So Io's continued volcanic activity is caused not by its own original heat, but by continued heating from its interaction with Jupiter.

Section 13.7

To summarize today's exercise, bring the *RedShift College Edition* screen back to the front. Click on the **Go** button in the navigation bar, select **Science of Astronomy**, **Worlds of Space**, **Hard Shells/Soft Centers**, and click on page **16**. Read and do the simulations and quizzes through page **33**.

Discussion

Mercury is by far the smallest of the terrestrial planets (though it is very dense), so it long ago lost most of its heat. Venus cooled just a little faster than Earth, so its crust is just a little thicker. It has volcanoes and mountain chains, but its crust is a little too thick to split into plates that could drift around like the continents of the Earth.

Mars, on the other hand, is only about half the Earth's size—just between Mercury and Venus. Because of its size, Mars had considerable early tectonic activity that soon dissipated as the heat was vented through the volcanoes and rift valleys. Mars's crust has now solidified completely, bringing to an end the volcanic activity that must once have been impressive. Our own Moon has a mass and density that just gave it the beginnings of tectonic activity before it also solidified rigidly.

Io is the only one of the bodies demonstrating volcanism or other tectonic features that does not fit the original heat/radioactive heat model for tectonic activity. As we have seen, though, Io's internal heat comes from a very different source—its parent planet, Jupiter.

Now we understand why some of the terrestrial bodies of our solar system have more tectonic features than others do. In a future exercise, we will try to discover why most of these active bodies are closer to the Sun and why the less active ones are farther out.

Conclusion

In a few paragraphs, summarize what you have learned today about the reasons for tectonic activity on the terrestrial planets and moons of our solar system. Include the different levels of tectonic activity (as evidenced by various members of the Solar System). Discuss the role of density in indicating tectonic activity and the processes of original heating and subsequent cooling in affecting the thickness of planetary crusts and the resultant amounts of tectonic activity.

13

Notes

Name: _____

Course: _____

Section: _____

Worlds in the Clouds

Introduction and Purpose

In the past several exercises, we have been examining the terrestrial planets. Today, we will venture beyond the orbit of Mars to look at the outer part of our solar system: the Jovian planets, or gas giants. As we saw in our exercise about the Earth, the atmosphere can be an important facet of a planetary body. On the Jovian planets, atmosphere makes up much of the planets' masses. These outer planets are truly *worlds in the clouds*.

In today's exercise, we will look at the features and characteristics of the Jovian planets. We will also make some brief comparisons between their atmospheres and the atmosphere of Earth.

Procedure

Section 14.1

Start *RedShift College Edition*. If the Contents frame is not open, click on the **Contents** button in the navigation bar to open it. Now click on the **Science of Astronomy** selection tab, and choose **Photos** to open the Photo Gallery window. In the Contents frame, click on **Solar System**, select **Jupiter**, and then click on **Jupiter** to open the Jupiter photo list. In the Jupiter photo list, scroll down to the image titled **[0738] Jupiter**. Enlarge this photo, and answer the following questions.

Q1 Back away from the monitor and squint at this photo to simulate the view through a telescope. How many large, dark stripes do you see going across the surface of Jupiter?

These stripes are the first thing you will notice about the planet when you see it in a telescope.

Q2 How many finer, less distinct stripes can you count?

Many of these bands are also visible in telescopes on nights when the air is very steady.

Q3 Study this photo for a few moments. List several different kinds of features you can see in the clouds.

14

Q4   In which direction do you think Jupiter is spinning in this picture, up and down or right and left?

Q5   Are both the major cloud belts in the same hemisphere of Jupiter?

Q6   Notice several small, dark orange-red spots toward the top of the planet. What shape are they?

Q7   Do to they appear to be raised up from the rest of the clouds or sunk in?

Q8   In the Photo Gallery Contents frame, select **Earth**, and then click on **Earth from space.** Scroll almost all the way to the bottom to the picture titled **[1413] Thunderstorms over Western Africa**, enlarge the photo, and compare it with the picture of Jupiter. Do you see any features in this photo that are similar in shape, perspective, and alignment to the small, dark orange-red spots in Jupiter's clouds?

Q9   The visible face of Jupiter is a lot easier to understand when we realize that it is just the tops of Jupiter's clouds that we are seeing. Since the face of Jupiter is illuminated by sunlight, do you think the darker areas are high points in the clouds or low spots?

Q10   Click on the **Previous** button in the navigation bar to go to **[1426] Thunderstorm systems over the Pacific Ocean**—another view of Earth's clouds. Here, it is easy to see how variations in altitude make darker and lighter regions in the clouds. The prevailing high-atmosphere winds have sculpted the cloud deck in the center of the picture into long, linear features much like those seen on Jupiter. Which way are the prevailing winds blowing in this picture?

Q11   Again, select and click on **Jupiter** in the Contents frame. This time, scroll down to the image titled **[0740] Jupiter's Great Red Spot**. Look first at the thumbnail image in the photo list. In this thumbnail photo, there are some features that look like they may be raised above the rest of the clouds (sort of like warts). In which part of the photo are they?

Q12   Enlarge the photo. Do these wart-like features resemble anything you have ever seen? They are eddy-currents much like the eddies in a river as the current sweeps past a rock or over some submerged feature. Here, the eddies are caused by the wind swirling away from the huge cyclonic feature in the upper right quadrant of the photo. This feature is, of course, the Great Red Spot. Just looking at this picture, which way do you think the Great Red Spot is rotating?

In the Atmospheric Activity column of the data table on page 164, write down the major atmospheric features on Jupiter.

Jupiter's Great Red Spot has been observed continuously for over 300 years. During that time, it has changed in color and size but has never disappeared. How could such a huge storm continue to swirl for so long? This question was not answered until the early 1980s, when the Voyager spacecraft flew past Jupiter.

To see what Voyager discovered, click on the **Go** button in the navigation bar, select **Science of Astronomy**, **Worlds of Space**, **Worlds in the Clouds**, and click on page **11**. Read and watch the simulations and movies through page **13**.

Q13 Are the clouds above and below the Great Red Spot moving in the same direction or in opposite directions?

Computer programmers set up a simulation of the clouds of Jupiter and entered the opposite direction wind speeds that you just observed. After a few years of simulated time, a swirling storm just like the Great Red Spot appeared near the equator of the simulated planet. The programmers let the simulation continue for many centuries of simulated time, and the cyclonic storm never disappeared.

Section 14.2

Now, click on the **Science of Astronomy** selection tab, and once again click on **Photos** to put the Photo Gallery contents back in the Contents frame. Select and click on **Neptune** in the **Solar System** section of the contents. Now scroll down to the picture titled **[2012] Neptune—dark spots and the "scooter"**. Enlarge this photo, and answer the following questions.

Q14 List any features on Neptune that are similar to features seen on Jupiter.

Q15 What do you think the white streaks are?

Q16 Now click the **Next** button in the navigation bar and see a close-up of some of those white streaks. Are these clouds higher than the surrounding atmosphere or lower? How can you tell?

14

Q17 Is the atmosphere of Neptune more turbulent than Jupiter's, or less?

In the Atmospheric Activity column of the data table on page 164, write down Neptune's major atmospheric features.

Section 14.3

In the Contents frame select and click on **Saturn**, and then scroll down to the picture titled **[3044] Saturn in August 1990 from HST**. Enlarge this photo, and answer the following questions.

Q18 How many large, dark cloud belts do you see on Saturn? (This feature is just barely visible in most amateur telescopes.)

Q19 How many small, fine cloud belts can you see?

Q20 How many major rings can you see easily? (In amateur telescopes, two major rings are usually visible.)

Q21 How many total rings (including subtle ring changes) can you count?

Q22 List any other features that remind you of either Jupiter or Neptune.

Q23 Click on the Thumbnail button and scroll down to the image titled **[3047] Storm on Saturn, December 1994 (HST)**. Enlarge the image, and list as many features as you can that remind you of Jupiter.

This storm on Saturn lasted for a few months. Such features on Saturn seem to be, on the whole, smaller and rarer than on Jupiter, but not nonexistent. In the Atmospheric Activity column of the data table on page 164, list Saturn's principal atmospheric features.

Section 14.4

In the Contents frame, select and click on **Uranus**, and then scroll down to the picture titled **[0855] Uranus in true and false colors**. Enlarge this picture, and answer the following questions.

Q24 In the left-hand view of Uranus, how many features can you detect?

Q25 Ignoring the tiny circles in the right-hand view (these are imaging artifacts and are not real), how many features can you detect in the false-color view that look like distinct clouds, swirls, or cloud belts?

In the Atmospheric Activity column of the data table on page 164, write down any atmospheric features you see on Uranus.

Q26 Rank the four Jovian planets in order of their atmospheric activity from most to least active.

1)   2)   3)   4)

Section 14.5

Rings are an important characteristic of the Jovian planets as well as atmospheres. In the Contents frame, select **Rings** from the **Saturn** section, and then scroll down to the picture titled **[0491] Under the rings (range 400,000 miles)**. Enlarge this picture, and answer the following questions.

Q27 How many rings can you count in this image?

Q28 Do you think you are seeing all the divisions there are in this region, or would there probably be even finer ring divisions if you got closer?

Q29 In the Contents frame, scroll down and select **Rings** from the **Uranus** section. In the photo list, find the image titled **[0863] Rings of Uranus, range 690,000 miles**. Enlarge this photo. How many rings can you count in this picture?

Q30 Now select **Rings** from the **Neptune** section of the Contents frame. In the photo list, find the image titled **[0469] Close-up of Neptune's rings**, and enlarge it. How many rings can you count in this thumbnail picture? (The rings are a lighter color.)

Q31 In the Contents frame, select **Rings** from the **Jupiter** section. In the photo list, find the image titled **[0741] Jupiter's ring**. Enlarge this photo, and see how many rings you can count.

Q32 Rank the four Jovian planets in order of their number of rings from most to least.

1)   2)   3)   4)

Section 14.6

Close the Photo Gallery window by clicking on its Close button. From the Information menu, open the **Find** window. In the **Find:** box, select **Planets**. One by one, find each planet listed in the data table, and double-

14

click on it. Then click on the **Information on...** link in the **List Links** box, and click **OK** to open the planet's information box.

Now click on the **More** button at the bottom of the information box to maximize the box. In the **Motion** section (which is initially open), find the length of a day on the planet and write this value in the Rotation column of the data table. Then click on the **Physical Data** tab, find the planet' s diameter, and record this value in the Diameter column of the data table. Finally, click on the **Rings & Moons** tab to find the number of rings and the number of moons for each planet and write these values in the Rings and Moons columns of the data table.

When you have entered all these values in the data table, close the information box and go on to the next planet.

Data Table

Jovian Planets					
Planet	Atmospheric Activity	Diameter (km)	Rotation (hours)	Moons	Rings
Jupiter					
Saturn					
Uranus					
Neptune					

Section 14.7

Q33 Think about the overall cloud patterns on the Jovian planets. Do the clouds seem to be randomly distributed as on Earth or congregated into linear jet stream patterns?

To understand why there is this large difference in cloud distribution, let's calculate the velocity of the atmosphere at the equator of each planet. The velocity at the equator of a planet is found by dividing the equatorial circumference of the planet by its rotation period. Thus, the formula is

Velocity = Circumference/Rotation Period

Circumference, however, is just the diameter of the planet times π, so the formula ends up as

Velocity = π * (Diameter)/(Rotation Period)

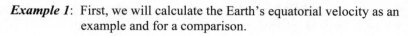

Example 1: First, we will calculate the Earth's equatorial velocity as an example and for a comparison.

$$Velocity_{Earth} = \pi * (Diameter_{Earth})/Rotation\ Period_{Earth}$$
$$Velocity = \pi * (12756\ km)/(24\ hours)$$
$$= 1670\ km/hr$$

This is almost exactly 1000 miles per hour! Follow the example to calculate the equatorial velocity of the atmospheres on the four Jovian planets.

Q34 Calculate the equatorial velocity on Jupiter.
$$Velocity = \pi * (Diameter)/(Rotation\ Period)$$
$$= \pi * (\quad)/(\quad)$$
$$= (\quad)\ km/hr$$
Round your answer off to the nearest whole number.

Q35 Calculate the equatorial velocity on Saturn.
$$Velocity = \pi * (Diameter)/(Rotation\ Period)$$
$$= \pi * (\quad)/(\quad)$$
$$= (\quad)\ km/hr$$
Round your answer off to the nearest whole number.

Q36 Calculate the equatorial velocity on Uranus.
$$Velocity = \pi * (Diameter)/(Rotation\ Period)$$
$$= \pi * (\quad)/(\quad)$$
$$= (\quad)\ km/hr$$
Round your answer off to the nearest whole number.

Q37 Calculate the equatorial velocity on Neptune.
$$Velocity = \pi * (Diameter)/(Rotation\ Period)$$
$$= \pi * (\quad)/(\quad)$$
$$= (\quad)\ km/hr$$
Round your answer off to the nearest whole number.

Q38 Rank the four Jovian planets in order of their equatorial velocities from highest to lowest.
1)   2)   3)   4)

Q39 How does your ranking in Q38 compare with your ranking in Q26?

14

Equatorial velocities help explain why cloud patterns on the Jovian planets are linear rather than random (as on Earth). However, they do not totally explain the amount of atmospheric turbulence on these planets.

Section 14.8

To discover the other factor involved in the turbulence of these gas giant planets, select *RedShift* 3/*RedShift* **college edition** from the window menu to reopen the *RedShift College Edition* window. Click on the selection tab (currently set for **Photos**), and select **Astronomy Lab**. Open the **Worlds Of Space** chapter in the Contents frame, and select the lab titled **Convection and turbulence**. In this interactive simulation, you will see how the rotation speed combines with the internal heat of the planet to affect the amount of turbulence in the clouds.

If you would like to do a more in-depth study of the Jovian planets, click on the Science of Astronomy menu button, and select **Science of Astronomy**. Select **Worlds in the Clouds** from the **Worlds of Space** chapter. Read, watch, and perform the simulations in this segment for a comprehensive summary of today's exercise.

Discussion

The Jovian planets are very different from the terrestrial ones. On a Jovian planet, the atmosphere makes up most of the planet. Today, we saw that these atmospheres exhibit many features observed on Earth, such as rotating storm systems and cloud banks at different altitudes. We also saw that the amount of atmospheric activity varies tremendously between these four planets. However, we noted some significant differences between the distribution of the cloud patterns on Earth (which are fairly random) and on the Jovian planets (which are mostly organized into linear cloud belts). Finally, we realized that at least part of the explanation for the differences in atmospheric behavior was due to the rotational velocity of the atmosphere near the planet's equator.

The Great Red Spot on Jupiter was a real puzzle for centuries. When the Voyager spacecraft flew past Jupiter in the early 1980s, it made measurements of the speed and direction of the clouds at various latitudes on Jupiter. We discovered that, as you might expect, Jupiter's equatorial winds blow at very high speeds. Further it was discovered that some of the adjacent linear cloud belts (jet streams) blow in opposite directions! Since that time, we have been able to demonstrate through computer simulations that such a cyclonic storm is a natural consequence of the interaction between these opposing high-speed winds.

Conclusion

In a couple of paragraphs, summarize what you have learned today about the Jovian planets. Compare their atmospheric characteristics, the number of moons for each, and their ring systems. Finally, suggest some possible experiments and measurements we might make in the future to help us better understand these worlds in the clouds.

14

Notes

Name: _____

Course: _____

Section: _____

Fire and Ice

Introduction and Purpose

In the previous exercise, we looked at the outer planets of our solar system, the gas giants. Today, we wish to focus on the moons of those great planets. Each of the four Jovian planets is itself a miniature solar system populated with all sorts of individual worlds—worlds of fire and worlds of ice.

As we examine these outer-system worlds, we will keep in mind the characteristics of the terrestrial planets and compare and contrast these worlds with those we have already studied. We will see many similar elements in many different settings.

Procedure
Section 15.1

Start *RedShift College Edition*, and then bring the *RedShift 3* screen to the front. From the Information menu, choose **Photo Gallery** to open the Photo Gallery window. If the Contents frame is not open, click on the **Contents** button in the navigation bar to open it. Click on **Solar System** in the Contents frame, then select **Jupiter**, and click on **Jovian System** to open the Jovian System photo list. Scroll down through the list of photos and enlarge the image titled **[0350] Jupiter's Galilean satellites**.

These are Jupiter's four large moons whose orbital periods we measured in the exercise "Galilean Merry-Go-Round." In this image, they are all shown at the same scale. Their names are (clockwise from the top left) Io, Europa, Callisto, and Ganymede. Study these four bodies. In the data table on page 176, write a brief summary of each moon's surface features noting your overall impression of each one. Then answer the following questions.

Q1 Which of the Galilean moons is largest?

Q2 Which is smallest?

Q3 Which one seems to have the fewest surface features?

Q4 Which one seems to have the most?

15

Section 15.2

In the Contents frame, click on **Io** in the **Jupiter** section. Scroll down through the Io photo list to the image titled **[0327] Io from 862,000 kilometers**, and enlarge it. Answer the following questions.

Q5   Comparing this picture of Io with the geological features of the terrestrial planets, would you say the circular features resulted from meteor impacts or volcanism?

Why?

Q6   Do you see many impact craters?

Q7   Look at the yellowish features. Of the three types of surface shaping forces we studied on the terrestrial planets—impacts, tectonics, and flow features—which would you guess caused these?

Q8   Look for other evidence that might suggest erosion from flowing material. Would you expect that fluid to be water, lava, or something else?

Click the **Next** button to see the culprit caught in the act. Io has numerous active volcanoes, all belching sulfur gas and spewing lava!

Q9   Click the Thumbnail icon and scroll back up to the image titled **[0753] Volcanic caldera** and lava flow on Io and enlarge it. Do you see any part of this surface that appears to be untouched by flowing lava and other ejected material?

Q10   How does your answer to the preceding question help explain why there aren't many craters on Io?

Add any items you wish to your summary of Io's surface features in the data table below.

Section 15.3

In the Contents frame, click on **Europa** in the **Jupiter** section. Scroll down through the Europa photo list to the image titled **[0747] Europa from 1.2 million miles** and enlarge it. Answer the following questions.

Q11   Do you see many impact craters in this picture?

Q12 How would you compare the color of Europa to that of the other moons?

Q13 Click on the Thumbnail button to return to the photo list. Scroll up to the image titled **[0748] Close-up of Europa** *(Be sure to get photo [0748])* and enlarge it. Describe this surface.

Q14 Does the surface seem to have any major high or low spots on it?

Q15 Look closely at the bottom left part of the picture. There is a bright, *sinuous* (curving back and forth) feature here. What kind of feature does it look like? (Is it raised up or sunken in?)

Q16 Does this surface look harder or softer than the surface of Io?

To see one of these cracks from a different angle, click the Thumbnail icon, and enlarge the image **[0322] Close-up of Europa**. You will see several of these cracks silhouetted on the left edge of the moon.

Q17 Let's summarize the characteristics of Europa. Fill in the following:
1) Color:
2) Surface terrain: (mostly flat or hilly)
3) Surface texture: (hard or soft)
4) Surface features:

Add any items you wish to your summary of Europa's surface features in the data table on page 176.

Q18 Think of all the common solar system materials we have previously discussed—water, lava, rock, iron, ice, etc. Which one of these meets all of the criteria listed in the previous question?

Now click the Thumbnail icon, and find and enlarge photo **[4013] Fractured Europa's icy crust**. Look carefully at the circular features. Have you ever seen a frozen pond that has cracked when heavy rocks were thrown onto its surface? The pond surface looks much like this photo. However, there are no rocks visible. What must be under this icy surface to allow rocks (or meteorites) to disappear?

Q19 A solid surface should fracture and leave a crater when hit by a meteorite; yet there are very few craters on Europa. There are clearly no volcanoes or erosion here to cover up or destroy the craters. Suppose the surface was made of ice. If you punch a hole through the ice on a frozen lake,

15

what will happen to the hole in a few hours?

Q20 Now click the Thumbnail icon, and find and enlarge photo **[4014] Europa ice-rafts**. Look for evidence of large sections of ice having been torn apart from each other. Has the surface refrozen?

Section 15.4

In the Contents frame, click on **Ganymede** in the **Jupiter** section. Scroll down through the Ganymede photo list to the image titled **[0745] Ganymede from 1.2 million kilometers**, and enlarge it.

Q21 Describe the amount of cratering in this photo.

Q22 Look at the larger craters. What do you observe extending away from them?

Q23 Does this finding indicate that these craters are younger than the rest of the surface or older?

Q24 Are the craters brighter than the rest of the surface or darker?

Q25 Given the common materials we've been discussing, which is most likely the composition of this bright, white material?

Return to the photo list, and enlarge **[0746] Ganymede from 151,800 miles** to answer the following questions.

Q26 Describe the amount and extent of features that look like flow features.

Q27 Are any of the features here more recent than the flows? (Look for craters on top of the flow features.)

Add any items you wish to your summary of Ganymede's surface features in the data table on page 176.

Section 15.5

Again in the Contents frame, click on **Callisto** in the **Jupiter** section. Scroll down through the Callisto photo list to the image titled **[0447] The Valhalla basin on Callisto**, and enlarge it.

Q28 What are the main features you see in this picture?

172

Q29 How are the craters on Callisto similar to those on Ganymede?

Q30 Do you see any sign of flow features or tectonic activity on Callisto?

Click the Thumbnail icon, and enlarge photo **[4010] Callisto crater-chain**. One of the interesting features recently discovered on Callisto is a long line of small craters that look like the path left by a rock skipping across a lake. This is exactly what would be expected if a meteor broke into numerous pieces shortly before impact and hit the surface at an extreme angle. In 1994, we got to witness such an event as comet Shoemaker-Levy 9 broke up while passing Jupiter and on the next orbit slammed one piece at a time into Jupiter's clouds. Similar chains have also been found on Earth's moon as well.

Add any items you wish to your summary of Callisto's surface features in the data table on page 176.

Section 15.6

The four Galilean moons of Jupiter are typical of most of the other moons of the outer planets. We will look briefly at two other moons to confirm that these features are common ones. In the Contents window, select **Saturn** and then click on **Enceladus**. Select the image titled **[0812] Enceladus** and enlarge it.

Q31 List the different kinds of surface features you see in this photo.

Q32 Which of the Galilean moons seems to have roughly the same mix of features as Enceladus?

Summarize Enceladus's surface features in the data table on page 176.

Q33 In the Contents frame, select **Saturn** and then click on **Dione**. Scroll down to the image titled **[0810] Dione** and enlarge it. Summarize the features you see on Dione in the data table on page 176. Notice the faint parallel streaks coming around the edge of the moon from the right. Do these streaks look like they are formed in the surface or sprayed across the top of it?

Some scientists have proposed that these streaks are the result of a comet splattering across the surface of Dione.

15

Section 15.7

Now, we are going to look at three really unique moons—Miranda, Titan and Triton. Again in the Contents frame, select **Uranus**, and then click on **Miranda**. Scroll down to the image titled **[0870] Mosaic of Miranda** and enlarge it. Summarize the features you see on Miranda in the data table on page 176. (You'd better write small!)

Q34 Are there any features seen on the other moons that you do *not* see on Miranda?

Q35 Describe any features on Miranda that you have not seen on any other moon.

Q36 Look at the bottom of the image. What does the feature just to the right of the bottom center look like?

The cliffs on the right of this feature are estimated to be around 20 km tall!

To gain a better understanding of just how incredibly tortured Miranda's surface is, click on the **Go** button in the navigation bar, select **Science of Astronomy**, **Worlds of Space**, **Moons, Asteroids and Comets**, and click on page **22**. Watch the movie and read the caption on this page. You may pause the movie at any point by clicking on the **Pause** button or replay it by clicking on the **Replay** button in the navigation bar.

Q37 Can you describe any combination of ordinary geological process that could produce such a hodgepodge of features?

Astronomers are still unsure about how Miranda came to have such a jumbled landscape. One of the most interesting theories suggests that Miranda was shattered by a collision with another planetoid or moon. The theory suggests that Miranda's pieces then slowly came back together under the influence of gravity. When they stuck themselves back together again, however, the original pieces were all mixed up, resulting in the crazy-quilt of strange shapes and odd angles we see today.

Q38 Click on the **Science of Astronomy** selection tab, and again select **Photos**. In the Contents frame, select **Titan** from the **Saturn** section. Scroll down to the image titled **[0807] Titan in true color** and enlarge it. Summarize the features you see on Titan in the data table on page 176. What are you seeing on the face of Titan?

Q39 Is there any other moon in our solar system whose surface you cannot see?

Q40 What would you guess about Titan's mass (and gravity) that would explain how it could keep an atmosphere?

Q41 Click Return, then click Contents and choose Triton from the Neptune section. Scroll down to the image titled **[2015] Triton** and enlarge it. Summarize the features you see on Triton in the data table on page 176. The bottom third of the moon is covered with pink frozen methane ice. Scattered here and there on the ice are a couple of long, black smudges. What do you think these might be? (Hint: Have you ever seen smoke blown by the wind away from a smokestack?)

Q42 Notice the three roughly circular, dark gray areas toward the upper right. We have features like this on our own Moon. What are they called? (No, they are not craters!)

Section 15.8

To finish up, we would like to know the densities and masses of these moons so that we might test the hypotheses that we constructed in the exercise "Fire and Brimstone" to explain surface features. Close the Photo Gallery window and then, from the Information menu, open the **Find** window. In the **Find:** box, select **Moons**.

One by one, find each moon listed in the data table and double-click on it. Then click on the **Information on...** link in the List Links box, and click **OK** to open the moon's information box.

Click on the **More** button at the bottom of the information box to maximize the box. Then click on the **Physical Data** tab to find the moon's density and mass, and record these values in the appropriate columns of the data table.

When you have entered all these values in the data table, close the information box, and go on to the next moon. When you have completed the data table, answer the following questions.

Data Table

15

Moons of the Outer Planets			
Moon	Surface Features	Mass (kg)	Density (g/cm³)
Io			
Europa			
Ganymede			
Callisto			
Titan			
Enceladus			
Dione			
Miranda			
Triton			

Q43 Look at the density of Ganymede. Remember that the density of ice is 0.9, the density of rock is 2.5, and the density of iron is 7.9. What percentage of the moon would you guess is made of:
1) Iron?
2) Rock?
3) Ice?

Q44 List the other moons with densities similar to or less than Ganymede's.

Q45 Leaving out Miranda, compare the surface features of these moons with Ganymede's.

Q46 Rank the four Galilean moons of Jupiter according to their densities from greatest to least.
1)   2)   3)   4)

Q47 Assuming that Europa has an ocean of liquid water beneath its surface of ice (that seems pretty certain, thanks to new information from the Galileo spacecraft), rank the four Galilean moons of Jupiter according to the amount of internal heat they seem to have (from greatest to least).
1)   2)   3)   4)

Q48 Do your rankings in the last two questions support the idea that denser bodies usually possess more internal heat and thus more tectonic activity and flow features?

Section 15.9

Internal heat is not the only factor that affects whether a surface exhibits flow features. If that were the only criteria, we would be shocked to find flow features on Saturn's moons, since their densities are only a little greater than water's. Whether or not a surface has flow features depends not only on the temperature, but on what *kinds* of materials are present.

Following is a chart of the temperatures at which various common compounds turn into liquids and then into solids. The second chart shows the temperatures at the surfaces of the moons of each planet. Study these charts, and then answer the following questions.

Maximum Liquid and Solid Temperatures—Common Solar System Materials					
State	Water	Carbon Dioxide	Ammonia	Methane	Nitrogen
Liquid	100 °C	—	-33 °C	-161°C	-196 °C
Solid	0 °C	-78.5 °C	-78 °C	-182 °C	-210 °C

Average Temperatures of the Outer Planets' Moons			
Jupiter	Saturn	Uranus	Neptune
-110 °C	-180 °C	-220 °C	-220 °C

Q49 Could any of these materials exist as a liquid on Jupiter's moons?

Q50 When the Sun heats the surface of a moon, the temperature can easily climb 30 to 50 degrees Celsius. If there are flow patterns on the surface of Neptune's moon Triton, what materials might flow?

Q51 Given the information about temperature variation in the last question, which of these materials might flow on Jupiter's moons?

Q52 If the Sun warmed the surface of Triton by as much as 50 degrees Celsius, what would happen to any nitrogen on or below its surface? Would it turn to a liquid or a gas?

This explains how Triton can have the geysers we have observed spouting above its surface. They are not made of water or lava, but of nitrogen.

15

Q53 Which two moons have the greatest mass (and therefore the greatest gravity)?

Q54 One of these has an atmosphere, but the other one doesn't. Look at the temperature charts above. What other factor is responsible for helping Titan hold its atmosphere?

Q55 If you were an interstellar traveler who accidentally stumbled onto a solar system with these moons as planets, which one of these nine would you visit first?

Why?

Discussion

The moons of the outer Solar System are unique worlds in their own rights. While many are cold and dead like our own Moon, many others are still active and changing. By measuring the densities of these moons, we have come to realize that there is a lot of water in the outer Solar System, just as there is water on the Earth. The only difference is its state. On Earth, all you have to do to get water is drill a hole and pump it out as a liquid. On most of the moons we studied today, you would have to dig mine shafts to look for this most precious substance. But that doesn't mean that these moons do not have erosion and flow features. As we also saw, many substances in our solar system are gases here on Earth but would be common liquids on some of the outer planets and moons.

We also verified that density is a key to understanding much about the makeup and activity of a planet or moon. By knowing a planet's density, we can make a pretty good guess about its constituent materials. We can also be pretty sure about the amount of internal heat present and thus about the types of surface features we will be likely to find there.

Conclusion

On a separate sheet of paper, describe the moons of the outer Solar System. First, summarize the typical features of an outer Solar System moon, using specific examples from your study today. Then discuss the unique moons we examined today—Io, Europa, Titan, Miranda, and Triton. Tell what is special about each of these moons, and then briefly describe the most probable reasons for each moon's unique characteristics.

Name: _____

Course: _____

Section: _____

The Big Tilt-A-Whirl

Introduction and Purpose

Our solar system is quite a place! It has terrestrial planets with solid surfaces, volcanoes, canyons, rivers, and oceans. It has rocky moons with craters, volcanoes, and melt valleys. It has great, gaseous worlds made of cloud canyons and jet streams.

Our solar system has a remarkable shape as well. The planets and moons move in a dizzying spin, each moon going around and around its planet which is in turn going around and around the Sun. It's just one big Tilt-A-Whirl! In today's exercise, we will conduct an overview of the motion of the Solar System. We will examine the shape of the Solar System, the planets' motions, and the motions of their moons. We will then consider the distribution of densities in the Solar System. Finally, we will try to put all the pieces together to come to some idea of how the Solar System might have formed in the first place.

Procedure

Section 16.1

Start *RedShift College Edition*, and then bring the *RedShift 3* screen to the front. In the Display menu, turn off everything except **Sun & Planets** and **Moons**. In the Location panel, click on the large Location button, and select **Space: heliocentric**. If the Lon (longitude) and Lat (latitude) boxes are not visible in the Location panel, click on the small icon at the upper right corner of the Location panel to expand it. Change the sky display as shown in the table below, and press the keyboard <Enter> key.

Lon (longitude)	180° 00′ 00″
Lat (latitude)	90° 00′ 00″
Dist (distance)	150 au
Step	15 days
Zoom	25

Now click on the small arrow to the right of the aim box, select **Sun**, and click on the **Lock Aim** button to lock the Sun in the center of the screen. You are now looking straight down on the North Pole of the Sun (90° latitude) from a distance of 150 au.

16

Section 16.2

Click on the Filters button, and select **Sun & Planets**. In the Sun & Planets filters window, click on the **Phases** box to remove the check mark. This action will show each planet as an illuminated disc even when it would normally be in shadow. Then click on the large **Labels** button, click on the large **Paths** button, and click **OK**. This will enable you to see the planets' paths and tell which planet is which. Now select **Ruler** from the Tools menu to open the Ruler tool.

Click repeatedly on the Forward Step button, and watch the planets orbit. Observe the direction of the motion of each of the inner planets. Write these directions in the Planet Revolution column of the data table on page 185 as either **CW** (clockwise) or **CC** (counterclockwise).

Continue to click the Forward Step button until all the planets have completed at least one orbit. (Mars will be the last to finish.)

Q1 In general, what shape are the inner planet orbits?

Q2 Are some rounder than others? (Measure the vertical diameter [the 12:00 position] of each orbit with the Ruler tool and compare it with the horizontal diameter [the 3:00 position].)

Q3 Which two are most elliptical (out-of-round)?

Q4 Do they all seem to be centered the same?

Q5 Which two are most off-center?

Section 16.3

Now, close the Ruler tool. Change the sky settings as shown in the table below and press the <Enter> key on your computer's keyboard to move out to where you can see the orbits of the outer planets.

Step	1 year
Zoom	1.3

Once again, repeatedly click the Forward Step button, and observe the direction of orbit of these planets. Record these directions in the Planet Revolution column of the data table as either **CW** or **CCW**.

Q6 Which of the planet orbits is different from the others?

Q7   What shape would you say these orbits are (besides the odd one)?

Q8   List two ways in which the odd orbit is different from the rest.

Section 16.4

Now, let's clean up the screen and change our viewpoint. First, open the **Sun & Planets** filters window and click on the large **Labels** button and the large **Orbits** button at the top to turn the names and orbits **Off**. Now click on the individual **Labels** and **Orbits** buttons for **Uranus**, **Neptune** and **Pluto**, and click **OK**. In the Location panel, change the coordinates as shown in the table below, and press the <Enter> key on the computer keyboard.

Lon	135° 00′ 00″
Lat	0° 00′ 00″

You *were* looking down on the Solar System from directly above the North Pole of the Sun. *Now* you are looking at it from the side. What do you notice about the orbits of Pluto and Neptune?

Since some of these orbits are tilted a lot and others are not tilted much at all, we wish to measure these tilts. (We call this tilt the *inclination* of the orbit.) Now select **Protractor** from the Tools menu to open the Protractor tool. To measure the inclination of each planet' s orbit, first click on the Sun, then pull a line straight out horizontally and click a second time. Finally, pull a line to the highest or lowest point in the planet' s orbit and click a third time. The resultant angle will appear in the protractor's angle box. Then answer the following questions.

Q9   Which of these planets has the largest inclination?

Q10   Are the inclinations of the other two similar to one another?

In the data table on page 185, record your measurement of each planet's orbital inclination.

Section 16.5

Reopen the Sun & Planets filters window, click **Off** the labels and orbits of **Uranus**, **Neptune** and **Pluto**, and click **On** the orbits and labels for **Jupiter** and **Saturn**. Then click **OK**. Change the Zoom to **5**, and press the <Enter> key on the computer keyboard so that you can see Jupiter's and

16

Saturn's orbits. Use the Protractor tool to measure the inclination of these orbits, and record them in the data table on page 185.

Q11   How do the angles of these two compare with those of Uranus, Neptune, and Pluto?

Section 16.6

Again, open the Sun & Planets filters window and click **Off** the orbits and labels of **Jupiter** and **Saturn** and click **On** the orbits and labels of **Mercury**, **Venus**, **Earth**, and **Mars**. Then click **OK**. Change the sky display as shown below and press the <Enter> key on the computer keyboard so that you can see the orbits of the inner planets.

Lon	55° 00′ 00″
Lat	00° 00′ 00″
Zoom	40

Again using the Protractor tool, answer the following questions.

Q12   Do any of these planets' orbits have a significant inclination?

Q13   How do the orbital inclinations of the inner planets compare with those of the outer ones?

In the data table on page 185, record your measurement of each planet's orbital inclination.

Section 16.7

Now, we will watch the rotation of each planet. Change the coordinates in the Location panel as shown below, and press the <Enter> key on the computer keyboard.

Lon	135° 00′ 00″
Lat	90° 00′ 00″
Dist	10 au

Open the Sun & Planets filters window, click **Off** the **Orbits** and **Labels** of the inner planets, and click **OK**. You will repeat the next section for each of the planets beginning, with Mercury.

Section 16.8

Unlock the current object (the Sun to begin with) by clicking on the **Lock Aim** button in the Aim panel. Click on the small arrow to the right of the Aim box, select **Planets/moons**, and then click on the planet you wish to watch (begin with **Mercury**). Then click on the **Lock Aim** button once more to lock the planet in the center of the screen. *(Don't forget this step!)*

Change the settings for each planet as shown here (begin with Mercury).

Planet	Step	Zoom
Mercury	**10 days**	**9999**
Venus	**10 days**	**9999**
Earth	**1 hour**	**9999**
Mars	**1 hour**	**9999**
Jupiter	**1 hour**	**3000**
Saturn	**1 hour**	**3000**
Neptune	**1 hour**	**3000**

*Uranus and Pluto have already been entered in the data table.

Now click the Play button in the Control Time panel, and watch the planet rotate. When you have determined in which direction it is rotating, record that value (either **CW** or **CC**) in the third column of the data table on page 185.

Click the Stop button and then repeat this section for **Venus**, **Earth**, **Mars**, **Jupiter**, **Saturn**, and **Neptune**. (Since Uranus and Pluto have no easily identifiable features to help you determine their rotation, we have already placed their data into the data table.)

Section 16.9

Now, we will turn our attention to the moons. Click on the Filters button, and select **Moons** to open the Moons filters window. We wish to click **On** the large **Image**, **Labels**, and **Orbits** buttons at the top of the window for all the moons. (The **Image** buttons will already be **On** for many of the moons. If so, do not turn them **Off**.) Begin with Earth's moon.

Now change the **Planet:** box to read **Mars**. Repeat the process of clicking **On** the large **Image**, **Labels**, and **Orbits** buttons at the top of the window. Change the **Planet:** box to **Jupiter**, and repeat the procedure. Do this again for **Saturn**, **Uranus**, and finally for **Neptune**. When you have turned **On** the images, labels, and orbits for all the moons of the planets

16

listed, click **OK** to record the changes. Your screen should still be set on Neptune, so begin by watching the revolution of Neptune's moons. (Make sure Neptune is still "**locked**.")

Section 16.10

Set the sky display as shown in the table below for each planet (beginning with Neptune), and press the keyboard <Enter> key.

Planet	Step	Zoom
Neptune	**1 day**	**300**
Uranus	**1 day**	**500**
Saturn	**1 day**	**500**
Jupiter	**1 day**	**500**
Mars	**1 hour**	**5000**
Earth	**1 day**	**500**

Click on the Play button and watch the direction of orbit of the moon (or moons). Record it as either **CW** (clockwise) or **CC** (counterclockwise) in the Moon Revolution column of the data table. Then click the Stop button.

In the Location panel, click on the **Lock Aim** button to unlock the current planet. Then click on the small arrow to the right of the Aim box, select **Planets/Moons**, and click on the next planet in the list. Click on the **Lock Aim** button again to lock the new planet in the center of the screen. Repeat this section for all the planets in the list.

Section 16.11

By now, you should have a pretty good idea of what's going on. Do you see a pattern developing? Let's confirm that pattern by watching the rotation of just a few of the moons. Open the Filters window, select **Moons**, change the Planet: box to **Earth**, and click the **Orbits** button to turn the orbits **Off**. Now change the Planet: box to **Jupiter**, click the large **Orbits** button **Off** there, too, and then click **OK**.

Of course, Earth has only one moon, so let's begin there. Click on the small arrow to the right of the Aim box, select **Planets/Moons**, choose **Earth's moon(s)**, and click on **Moon**. Make sure the Moon is "**locked**," and then change the display settings as shown in the next table.

184

Step	1 day
Lon	135° 00′ 00″
Lat	90° 00′ 00″
Dist	1.000 m km (million kilometers)
Zoom	9999

Click the Play button and watch the rotation. (If the Earth gets in the way, just wait until it moves off the screen.) When you can tell in which direction the Moon is rotating, record the answer in the appropriate space in the Moon Rotation column of the data table and then click the Stop button.

Section 16.12

Now set the Step to **6 hours** for the moons of Jupiter and press the keyboard <Enter> key. Again click on the small arrow to the right of the Aim box, select **Planets/Moons**, choose **Jupiter's moon(s)**, and then click on **Io**. Again, make sure Io is "**locked**."

Click on the Play button, and watch the rotation. (If Jupiter gets in the way, wait until it moves off the screen.) When you can tell in which direction the moon is rotating, record the answer in Jupiter's row of the Moon Rotation column, and click the Stop button. Repeat this procedure to check the remaining Galilean moons, Europa, Ganymede, and Callisto.

Data Table

The Planets						
Planet	Planet Revolution	Inclination of Orbit	Planet Rotation	Moon Revolution	Moon Rotation	Planet Density
Mercury						
Venus						
Earth						
Mars						
Jupiter						
Saturn						
Uranus			CW			
Neptune						
Pluto			CC			

Q14 In which direction is our solar system spinning?

16

Q15 Which planets are not spinning in the same direction?

Q16 Which planet is spinning on its side?

Q17 Which planet's orbit does not seem to belong with the rest?

Q18 In general, what can you say about the spins of the planets and moons of the Solar System?

Q19 Do the spins of the Solar System bodies seem to indicate that most of them formed together at the same time in the same way, or separately under different conditions?

Q20 How could you tell if a moon or a planet did not form with all the rest?

Q21 If a moon or planet did not form with all the others in its vicinity, suggest a theory to explain how it came to be here.

Q22 Uranus is spinning on its side. Do you think it was tipped before or after its moons formed?

Why?

Section 16.13

From the Information menu, open the **Find** window. In the **Find:** box select **Planets**. One by one, find each planet listed in the data table, and double click on it. Then click on the **Information on...** link in the List Links box, and click **OK** to open the planet's information box.

Click on the **More** button at the bottom of the information box to maximize the box, and then select the **Physical Data** tab. In this window, find the density of each planet, and enter it in the last column of the data table. When you have entered the density in the data table, close the information box and go on to the next planet. When you have completed the data table, answer the next question.

Q23 Look at the densities of the planets. What trend can you find in densities from the inner planets to the outer ones?

It's tempting to think that gravity held the more dense material close to the Sun while the less dense material flew further out as the Solar System began to spin. However, the real explanation for this trend of densities is the temperature at various distances from the Sun.

Temperature is just a measure of the speed of the molecules of a substance, so a higher temperature means a higher molecular speed. If a molecule of gas is hot enough (that is, going fast enough), it will escape from the gravity of a small planet or moon. However, if the planet or moon has a very large gravity (like Jupiter) and the speed of the molecules at the top of its atmosphere is very slow (at too low a temperature), the molecules cannot escape and must remain as a part of the planet or moon's atmosphere.

To see how this works, select **RedShift 3/RedShift College Edition** from the **Window** menu to switch back to the RSCE window. Now, click on the **Science of Astronomy** Tab and select **Astronomy Lab**. In the table of contents that appears on the left, click on the folder titled **Worlds of Space**, and select the lab titled **Primary Atmospheres**.

In this experiment, you can vary the planet's distance from the Sun and its mass. Try various combinations of these two factors to see how a large planet (like Jupiter) could hold onto lightweight gases like hydrogen and helium, while smaller planets that were nearer to the Sun could not keep their primary atmospheres. Use the simulation to determine the answers to the following questions.

Q24 What is the smallest mass of planet that could hold hydrogen and helium at the distance of the Earth to the Sun?

Q25 What is the smallest mass of planet that could hold hydrogen and helium at Mercury's distance from the Sun?

As you saw, the small inner planets could not hold on to large amounts of hydrogen and helium because they were too small and too hot. The large outer planets, on the other hand, were cool enough (and large enough) to hold on to the gases in their own neighborhoods and to capture the gases that were swept away from the inner planets when the Sun began to shine.

16

Section 16.14

Now we would like to explain the formation of a solar system where everything is spinning in pretty much the same direction and where most of the mass is distributed in a flat plane—like a dinner plate or a Frisbee.

If we can explain how such a system might have formed, then it will be easy to understand the distribution of densities. As we have seen in previous exercises, once we understand the distribution of densities, it is easy to explain the varying amounts of tectonic activity and the natures of the inner and outer Solar System planets.

Q26   One simple example will help us in our quest for a theory. Have you ever seen a pizza chef making pizzas? He starts with a well-kneaded ball of dough. What shape is it to start with?

Q27   What does a show-off pizza chef do to change the shape from a ball to a flat, round crust?

Great pizza chefs spin the dough to make it flatten out into a lovely, smooth pizza crust. The spinning action slings the less well-attached parts of the dough to the edge and pulls the lumpy dough in the center outward and toward the plane of the pizza crust at the same time. This feeds more and more dough outward into the flat regions and reduces the size of the ball in the center.

Q28   In pizza-making, two factors affect the dough—its natural stickiness caused by the gluten in the flour, and the inertia of the spin. What causes the dough to want to fly apart?

Q29   What *keeps* the dough from going to pieces and flying apart?

To see how this model helps explain the shape of the Solar System, insert the Extras CD in your computer's CD-ROM drive. When the small Start screen opens, click **Start** to open the Extras title page. In the Extras title page, click on **The Story of the Universe** button (you may have to scroll down), and in The Story of the Universe contents list, scroll down and select the item titled **History of the Solar System**. Watch this narrated animation, and then answer the following questions.

Q30   In the early Solar System, there were two factors doing the same things as the gluten and the inertia of the pizza dough. An initial impetus (perhaps a supernova shock wave) passed through the solar nebula and pushed the material toward a common center. The shock wave continued moving on, so what would happen to the center of the condensation as well?

Q31 As the outlying mass fell toward the mass center, the center was moving. By the time the outlying mass got to where the center was, the center was no longer there. The outlying mass had to curve in its path to try to catch up with it. What would be the result of such a continuously curving path?

Q32 Your answer to the last question explains the rotation of the Solar System. What held the Solar System materials together and kept it from flying apart once it started moving?

Section 16.15

To summarize what you have learned today, close the Extras window, and reinsert the *RedShift College Edition* CD in the CD-ROM drive. After the drive has spun and recognized the new CD, select ***RedShift 3/RedShift college edition*** from the main *RedShift 3* Window menu to reopen the *RedShift College Edition* window. Open the Contents frame, and select **Other Solar Systems** from the **Worlds of Space** chapter of the **Science of Astronomy**. In this segment, read and watch through page **16** for a good review of today's exercise.

Discussion

Now, we have a pretty good idea of how our solar system formed. The initial rotation produced by the motion of the center of the proto-solar nebula gradually gathered momentum. Gravity was pulling the heavier materials toward the center and it was heating up and spinning faster, while the lighter materials stayed in the outer, cooler regions.

What began as a more or less spherical clump of gas and dust slowly began to flatten out like a lump of pizza dough as the rotation increased. Just as the stickiness of the dough *pulls* material out of the central dough ball into the spinning disk, so gravity *pulled* the material from above and below the rotation plane into the disc of the ecliptic. The only escapees from this process were icy objects so far away that they were not drawn in until much later—the comets.

Conclusion

In the space below, summarize the process of the formation of the Solar System. Mention the rotation and revolution of the Solar System's bodies and its distribution of densities. Explain how the rotation helps explain the densities and how the densities help explain the surface characteristics of the planets and moons. Finally, explain why the outer planets are so large despite their small densities.

III
Stars

Contents

Name: _____

Course: _____

Section: _____

The Sun

Introduction and Purpose

The purpose of today's lab is to familiarize you with our Sun. The Sun is the nearest star to the Earth, and as such is our best example of what a star is like. In today's lab, you will get an opportunity to explore the database contained in the *RedShift* software to learn about the various characteristics of the Sun. Rather than watch a simulation, you will be asked to look closely and carefully at several photos of the Sun and to record the interesting features and characteristics you observe in each. Along the way, you should also make notes of information regarding the solar cycle and how it affects life on our Earth.

Procedure

Section 17.1

Start *RedShift College Edition*. (If you're not sure how to do this, refer to Appendix A.) Click on the large **Contents** button in the navigation bar at the top of the screen to open (or close) the Contents frame. The Contents frame resides at the left edge of the window when it is open. Click on the **Contents** button repeatedly to see how it operates. With the Contents frame open, click on the **Science of Astronomy** selection tab at the bottom of the navigation bar, and select **Photos** from the drop-down menu to open the Photo Gallery and to put the Photo Gallery contents list in the Contents frame. In the Contents frame, click on **Solar System** to open the Solar System folder, and select **Sun** from the list of topics. When the list of Sun photos appears in the main part of the window, use the scroll bar to move down to the image titled **[0562] The visible surface of the Sun.**

Click on the small Magnify icon to enlarge the photo, and spend a minute or so carefully observing this image. In the following data table, make a note of all of the major features you see. Include such things as spots, variances in coloration (even tiny or slight ones), and lighter or darker areas. When you have finished, click the Thumbnail icon to return to the photo list.

Data Table 1

Features in the visible light image	
Central Part of the Sun	Limb (Edges) of the Sun
1. Orang — yellow center	1. Black
2. some black spots	2. Fades slowly
3.	3.
4.	4.

Section 17.2

Scroll through the photos to the one titled **[1167] The Sun photographed in red hydrogen-alpha light**. Click on the Magnify icon to enlarge this image, and study it carefully. In the data table below, make a note of all of the major features you see. Include such things as spots, variances in coloration (even tiny or slight ones), and lighter or darker areas.

Data Table 2

Features in the hydrogen-alpha light image	
Central Part of the Sun	Limb (Edges) of the Sun
1. same color	1. same color
2. rough surface	2.
3. whit spots	3.
4. crak in surface	4.

Section 17.3

Click the Thumbnail icon to return to the photo list. Scroll through the photos to the one titled **[2004] Total eclipse of the Sun in 1970**. As before, enlarge this image and examine it thoroughly. In Data Table 3, list all the major features you see, including variances in coloration (even tiny or slight ones), asymmetric shapes, streamers, and lighter or darker areas.

Data Table 3

Features in the solar corona image	
1. dark green corna	3.
2. black sphere	4. white sphere under corna

Section 17.4

Click the Thumbnail icon to continue. Scroll through the photos to the one titled **[1155] A solar flare in hydrogen-alpha light**. Enlarge this image and carefully observe it. In the data table below, note all the major features you see. Include such things as spots, variances in coloration (even tiny or slight ones), and lighter or darker areas.

The flare in this image does not look as impressive as it really is. To get a better feeling for what these flares are doing, click on the large **Go** button in the navigation bar. From the successive drop-down menus, select **Science of Astronomy**, **Stars**, and **The Sun**, and click on page **15**. Watch this X-ray movie of the rotating Sun. The bright spots are flares. As you can see, some of these flares are putting off thousands of times more energy than the rest of the Sun's surface.

Data Table 4

Features in the hydrogen-alpha flare image	
Central Part of the Sun	Limb (Edges) of the Sun
1. mixed	1. mixed
2. darker center	2. red corona
3.	3. Bright spots
4.	4. some red haze

Section 17.5

Return to the Photo Gallery by clicking on the **Science of Astronomy** selection tab and selecting **Photos**. Make sure **Sun** is still selected from the **Solar System** section. Scroll through the Sun photo list to the one titled **[0410] A gargantuan solar prominence**. As before, enlarge the image, and note all the major features you see, recording them in the data table. Include such things as spots, variances in coloration, and lighter or darker areas.

Data Table 5

Features in the prominence image		
Central Part of the Sun	Limb (Edges) of the Sun	
1. *Red and white spots*	1. *large flare*	
2. *dark spots*	2. *black corona*	
3. —	3. *several white spots*	
4. —	4.	

(handwritten entries)

To get a better idea of just how incredible these prominences are, let's watch one in motion. Click on the large **Go** button in the navigation bar. Select **Science of Astronomy**, **Stars**, and **The Sun**, and click on page **10**. Watch this actual time-lapse movie of a solar prominence leaping out from the surface of the Sun.

Section 17.6

Now, let's put these images and movies into perspective. Click the large **Prev** (Previous) button in the navigation bar successive times until you are on page **7**. Watch, read, and perform the simulations through page **28**. Then answer the following questions.

Q1 What is a sunspot?
    A dead spot in the sun.

Q2 How many years does one complete sunspot cycle take?
    11

Q3 Do the sunspots always appear at the same latitudes on the Sun?
    no

Q4 Has the sunspot activity always been very constant? Give some reasons for your answer.
    yes

Q5 Tell how scientists have been able to trace sunspot activity back several thousand years.

Q6 Look back through the photos you observed in today's exercise. On which of the Photo Gallery image(s) do sunspots appear?
    11 6 7

Q7 On which image(s) do prominences appear?
    11 5 5

196

Web Intrigues

(If you do not have Internet access, you may skip this section.)

Log on to the Internet and open your web browser. Type in the following URL: *http://www.planetscapes.com/solar/eng/homepage.htm* to open the "Views of the Solar System" page. Once this page has opened, click on **Sun**. This web site has lots of great photos and movies about the Sun. Of special interest to future teachers are the Educator's Guide sections that describe experiments and simulations that can be done easily in most classrooms to illustrate difficult concepts. Find the "Educator's Guide to Convection," and go through it. (If this site is not available, look up "Sun" in a good search engine to answer the following questions.)

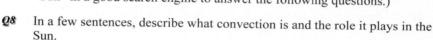

Q8 In a few sentences, describe what convection is and the role it plays in the Sun.

Click the **Back** button in your browser window to return to the main Sun page. Go through the pictures and read the captions. (If you have a fairly fast Internet connection, load and watch one or more of the SOHO movies—they are fantastic!)

Q9 Briefly describe the image (or movie) that most impressed you on the Sun page.

Section 17.7

Insert the Extras CD into your computer's CD-ROM drive. When the small Start screen opens, click **Start** to open the Extras title page. Here, click on **The Story of the Universe** and, in The Story of the Universe contents, select the item titled **The Sun**. A movie window will open, and a narrated animation describing the Sun will begin playing. This animation is an excellent summary of this exercise.

To replay any part of the animation, move the cursor over the horizontal position bar (at the bottom of the window) to the approximate position of the part you want to see again, and click.

(For example, click at the left end of the bar to begin all over, the center to begin watching near the middle, etc.) Watch this narrated animation and use the information there to answer the following questions.

17

Q10 What is the visible part of the Sun called?

Q11 What are the two parts of the Sun that are visible only during total eclipses?

Q12 What two gases (in order of their percentages) make up most of the Sun's mass?

Q13 How is a flare different from a prominence?

Q14 What is the temperature in the center of the Sun?

Q15 How much hydrogen does the Sun destroy every second?

Q16 What is a *granulation*?

Q17 What is a *spicule*?

Q18 Why do auroras occur mostly near the north and south poles of the Earth?

Q19 How long does it take for the Sun to rotate on its axis?

Discussion

The Sun is an incredibly violent place. It is powered at its core by thermonuclear reactions where the incredible temperature (over 15,000,000 degrees Kelvin) and enormous pressure forces hydrogen atoms to collapse into one another, fusing together to become helium. Fusion releases a tremendous amount of energy that then travels slowly from the core outward toward the surface of the Sun. Much of this energy transfer is accomplished through convection, the same mechanism that causes bubbles of hot oatmeal when it begins to thicken to rise to the top of the pot. As the bubbles of superheated gas arrive at the surface of the Sun, they create the various features we observe. The hotter and cooler areas create the granular appearance and the *faculae.*

When there is a strong magnetic field in a certain area, the convection is often repressed or diverted. The repressed areas are then cooler and appear as dark *sunspots*, while the areas to which the convection is diverted often erupt as *flares,* creating *filaments* and *prominences.* Some of the gases reach the surface at such high speeds that they escape the gravity of the Sun altogether and form the *corona* as they speed away from the surface of the Sun. Because the overall light output of the Sun is so great, many of these features are visible only when the Sun is eclipsed or when all the light is filtered out except that coming from hydrogen atoms (the hydrogen-alpha emission).

Conclusion

When we look at the Sun in different ways, such as during an eclipse or in hydrogen-alpha light, we can begin to put together a fairly complete picture of how its surface behaves. Write a couple of short paragraphs describing the various features that can be observed on the Sun. Explain which ones need special conditions to observe them and what they tell us about the Sun's nature. Include a short discussion on the effects of solar activity (such as sunspots and flares) on the Earth and its atmosphere.

17

Notes

Name: _____

Course: _____

Section: _____

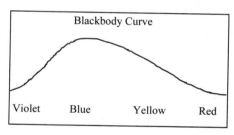

Star Colors

Introduction and Purpose

Anyone who has been outside on a frosty winter night in the country, the desert, or the mountains knows how awesome the night sky can be when there is nothing to block our view of the stars. Some stars are incredibly bright, some are different colors, and some are tiny pinpoints in the black velvet of space. In this exercise, you will discover some of the basic properties of the stars we can see—what colors they are, how hot they are, and what colors are most common.

Before you do today's exercise, you need to know one very basic thing: *99.9% of the light that comes from a star is produced by the heat of that star.* When we pass a star's light through a ***spectroscope*** (look this term up in the dictionary later), we find that this light is spread over a broad range of colors. We call this the star's spectrum. Most stellar spectra look just like the rainbows we are so familiar with here on Earth. They all have the characteristic shape (called the blackbody curve) shown below.

The horizontal axis of the graph represents the colors of the rainbow with red on the right, green in the middle, and violet on the left. However, instead of being marked off in colors, the axis is usually marked off in the wavelengths of the light that correspond to these colors.

The intensity of each color in the star's spectrum is indicated by the height of the curve at that color's wavelength. Thus, the tallest point on the curve represents the most intense color coming from the star. The actual wavelength where this occurs varies from star to star. The purpose of today's exercise is to discover what colors the bright stars are, which temperatures correspond to those colors, and which colors are most common. Along the way, we will get some idea of just how many stars there might be.

201

Procedure
Section 18.1

Start *RedShift College Edition*, and then bring the *RedShift 3* screen to the front. (If you have just started with this book or are not sure how to bring up a screen, please see Appendix A.) Once you have figured out how to go to the *RedShift 3* screen and have set your location and other basic parameters, save your personal sky settings by following the instructions in Appendix B. Once you have created a personal settings file, open it using the **Open Settings** option from the File menu.

Now that you have set the basic sky parameters, open the Display menu and turn off all items except **Stars**, **Deep Sky**, and **Constellations**. (If any other items have check marks beside them, click on them to remove the check marks. Continue opening the Display menu and removing check marks until only the items titled **Stars**, **Deep Sky** and **Constellations**, have check marks beside them.)

Now, click on the Filters button, and select **Stars** from the drop-down menu to open the Stars filters menu. For today's exercise, we want to emphasize the color of the stars. In the lower right corner of the Filters window, you will find a slider control marked **Saturation**. Click on this slider, and drag it all the way to the right so that the Saturation box reads **100**. Now, click the **OK** button to close the Filters window. You should see now that most of the stars have very pronounced colors.

Section 18.2

Turn your attention to the Aim panel. Click on the small arrow to the right of the Aim box, select **Deep Sky**, and click on the **Hyades**. The sky should rotate to center your view on a V-shaped group of stars. This big V in the sky is the face of Taurus the bull, and the cluster that forms it is known as the Hyades. Now click on the small arrow to the right of the Zoom box (at the bottom of the Aim panel). Select **3.5** from the drop-down menu to increase the magnification. This beautiful open star cluster is full of brightly colored stars, with more than its share of red, orange and yellow stars.

One by one, click on some of the red and orange stars to open their name tags. Record the names (or abbreviations) shown on the name tag in the data table on the next page in the Red and Orange section. Now click on the name tag to open the star' s information box. In the information box, click on the **More** button to completely expand the box. Finally, click on the tab titled **Color and Luminosity** to display that information in the box. In the middle of the box you'll find the words **Spectral type**. Record the letter and number at the beginning of the spectral type in the Spectral Type column of the data table below. (For instance, a spectral type of B2IV-V would be recorded as a B2). Record the color listed just below the spectral type in the Color column of the data table below. Just below the spectral type information, you will find three boxes labeled **V, B**, and **(B - V)**. Record the B-V index number in the B-V column of the data table. Finally, estimate the star's temperature from the graph at the right of the window and write that in the temperature column. Now, click the **Close** button to close this star's information box, and then repeat the entire process for other red and orange stars until you have filled all the blank lines in the Red and Orange Stars data table.

Data Table 1

Red and Orange Stars (Hyades)				
Star name	Spectral type	Color	B-V Index	Temperature

Section 18.3

Now let's move just a little way through the sky to the constellation of Orion to find some blue and white stars. To rotate the sky to center on Orion, click on the small arrow to the right of the Aim box, select **Deep Sky**, and click on **Orion Nebula**. In this area, concentrate on the blue and white stars.

As you did with the red and orange stars in the Hyades, choose a blue or blue-white star, and click on it to open its name tag. Record the name (or abbreviation) shown on the name tag in Data Table 2 below in the Blue and White section. Now click on the name tag to open the star's information box. In the information box, click on the **More** button, and then click on the tab titled **Color and Luminosity**. Again, record the letter and number at the beginning of the spectral type in the Spectral Type

18

column. Then record the color shown just below the spectral type in the Color column, and record the B-V index number in the B-V column of the data table. Finally, estimate the star's temperature from the graph at the right of the window, and write that in the temperature column. Click the **Close** button to close this star's information box. Then, repeat the entire process for other blue and white stars until you have filled all the blank lines in the Blue and White Stars data table.

Data Table 2

Blue and White Stars (Orion)				
Star name	Spectral type	Color	B-V Index	Temperature

Now answer the following questions.

Q1 Look at the data you recorded in the first two data tables. On the basis of your observations, which color (or colors) of star seem(s) to be the hottest?

Q2 Which color (or colors) of star seem(s) to be the coolest?

Q3 Based on your observations, what color would a star with a B-V index of −0.20 be?

Q4 Approximately what color would a star with a B-V index of +1.50 be?

Q5 With those extremes, what color would a star with a B-V index of +0.40 be?

Q6 Given the temperatures of the stars in your survey, make a rough estimate of the temperature of a star with a B-V index of +0.40.

To verify your answer, click on the small arrow to the right of the Aim box, select **Stars**, and click on **Procyon** to center that bright wintertime star. Click on Procyon and on its name tag to open its information box and check its temperature, color, and spectral type. Adjust your answer if you need to.

Section 18.4

From the Information menu, choose **Photo Gallery**. This will open the Photo Gallery window. In the Photo Gallery Contents frame, click on the section titled **The Galaxy**, and then select **Open clusters** to open a list of photos of open star clusters. When the open clusters photo list opens, use the scroll bar at the right side of the window to scroll down to the photo titled **[0239] A young open cluster in Carina, NGC 3293**.

Click on the small Magnify icon on the picture to enlarge the photo. Notice the colors of the stars. What are the colors of the brightest stars? Now, divide the image into quarters, and, using only the lower left-hand quarter, count the stars that appear blue, those that appear white or cream color, those that appear yellow or orange, and those that appear red or pink. (Don't worry if you miss a few.) Multiply these counts by 4 to get a rough total for the whole picture, and record these counts in the appropriate boxes in the data section of the lab. Finally, rank the colors as dim (3), average (2) or bright (1) in the Rating column. (More than one color may be ranked 1.)

Data Table 3

NGC 3293							
Color	Stars (¼ screen)	Stars (Total)	Rating	Color	Stars (¼ screen)	Stars (Total)	Rating
Blue				**Yellow or Orange**			
White				**Red or Pink**			

Section 18.5

Now click the Thumbnail icon to return to the photo list. Use the scroll bar to scroll back up to the image titled **[0233] Dust cloud and open cluster NGC 6520**. Again, enlarge the image and note the colors of the various stars.

In this image, there are far too many stars to count. Note which kinds of stars are the brightest: blue, white, yellow/orange, or red/pink. Now,

18

choose a region that is representative of the whole picture, and count the number of stars (of all colors) in one square centimeter. (Use a ruler to estimate this area.) Measure the dimensions of the whole image in centimeters. Write the dimensions and the number of stars in the data table below. Now, calculate the area of the screen by multiplying its length by its width, and write that figure in the appropriate space in the data table below. Finally, multiply this area by the rough count that you made for the one square centimeter, and record this total in the data table.

Data Table 4

NGC 6520			
Color	Rating	Color	Rating
Blue		Yellow/Orange	
White		Red	

NGC 6520					
Photo Dimensions	Length (squares)	Width (squares)	Total Area (squares)	Number of stars per square	Total number of stars

Section 18.6

Open the Window menu, and select *RedShift* 3/*RedShift* **college edition** to switch to the *RedShift College Edition* window. In this window, click on the selection tab in the navigation bar (currently set to **Photos**), and from the menu that drops down, select **Science of Astronomy**. In the Contents frame, click on the **Types of stars** segment in the **Stars** chapter. Read, listen, and perform the experiments as you go through this segment to page **11**.

When you have gone through this segment, use the information you learned and the observations you recorded in the data table to answer the following questions.

Q7 On the basis of the star counts and ratios you estimated, which star color (or colors) seem(s) the most common?

Q8 On the basis of your observations of both photos, which two colors of stars are the brightest?

Q9 Are the brightest stars a little brighter than most of the others or a lot brighter?

Q10 If two stars were the same size, but one was hotter than the other, would you expect the hotter one to be brighter than, dimmer than, or the same brightness as the cooler one?

Q11 Q8 implied that two colors were brighter than all the rest. Which of these two colors *should* be brighter, based on your answer to Q10?

Q12 Which color of star seems to be the faintest in the photos?

Q13 Based on your answer to Q10, does this make sense?

Q14 If some stars are basically brighter, is there any way to know a star's distance just by looking at its brightness?

Q15 On the basis of your estimates of the percentages of stars of each color in Data Table 3, are most of the bright stars the same color as most of the average stars, or are they actually the rare and unusual ones?

Q16 Were you surprised by your estimate of the total number of stars in these pictures?

Q17 On the basis of your observations, would you say that *most* of the brightest stars are
A)   The hottest ones?
B)   The moderately hot ones?
C)   The coolest ones?

Q18 On the basis of your observations, would you say that *most* of the dim stars are
A)   The hottest ones?
B)   The moderately hot ones?
C)   The coolest ones?

18

Q19 What color of star sometimes seems to disobey the general rule you set up in Q17?

Q20 Did you notice any stars that seemed to disobey the general rule you set up in Q18?

Section 18.7

Based on your observations, write the approximate color of each spectral type listed in the table below.

Temperature (K)	>25,000	11,000-25,000	7,500-11,000	6,000-7,500	5,000-6,000	3,500-5,000	<3,500
Spectral Class	O	B	A	F	G	K	M
Color							

Discussion

In today's lab, we have observed that stars come in many different colors that actually indicate many different temperatures. We saw that, in general, the very bright stars are also very hot. We saw that for the most part, the dim stars were also very cool. Contrary to our intuition, however, a few moderately cool stars are nonetheless very bright. We also saw that the types of stars that are brightest are not necessarily the most common.

Conclusion

On a separate sheet of paper, summarize what you have learned today about the different kinds of stars that we can see in the sky. Mention their temperatures (and colors), which colors are most common, which are brightest, and which are dimmest.

Based on your estimates of the numbers of stars in each photo, discuss the possible total number of stars in the sky. Finally, based on the differences between the two photos you observed, remark on the concentration of stars in different parts of the sky.

Optional Observing Project

Do the observation exercise "Star Colors" in Appendix C.

Name: _____

Course: _____

Section: _____

The Distances to the Stars

Introduction and Purpose

As you look at the stars twinkling overhead, have you ever wondered just how far away they are? Sometimes they seem to be just beyond your reach. Yet, if you have read much, you know that they are very, very far away. In today's lab, you will discover just how astronomers measure the distances to the nearby stars. Then you will use the data that astronomers have collected in the past two hundred years to do your own calculations of the distances to the nearby stars.

The only method that we have to *directly* measure the distance to any object out in space is called ***parallax***. To understand how the parallax method works, do this simple experiment. Hold up your index finger about 10 inches in front of your face. Close your left eye, and move your hand to line up your finger with something on the other side of the room. Now close your right eye, and open your left eye without moving your finger. What did your finger seem to do? This apparent motion of your finger is **parallax**. It is easy to understand if you draw a diagram.

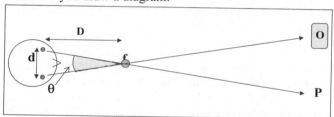

Your right eye is lined up with both your finger, **f**, and the distant object, **O**. Because of the separation of your eyes, **d**, however, your left eye sees your finger lined up with some other point, **P**. The angle, **θ**, is called the parallax angle, or just the parallax. The angle, **θ**, depends on the separation between your eyes, **d**, and the distance, **D**, from your eyes to your finger. The equation relating them is

$$D = (2.06 \times 10^5)*d/\theta \qquad \textbf{\textit{Equation 1}}$$

* *The 2.06 × 10⁵ converts the numbers to the proper units for our measurements.*

Now repeat the experiment with your finger at arm's length. What differences do you observe from the first trial? It is easy to extrapolate this to even larger distances. The farther away your finger is, the less it seems to move. Eventually, if you could move your finger far enough away from your face, the amount of the parallax would become so small that your eyes would not be able to detect it.

In astronomy, we use this method by observing stars while the Earth is at one side of its orbit around the Sun and then observing them again three months later when the Earth is at the center of its orbit.

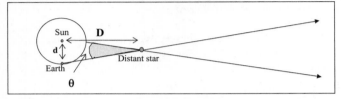

This provides two viewpoints separated by a distance just like our two eyes. The distance between our eyes is quite small, but the radius of the Earth's orbit is 149.6 million km, or 1.58×10^{-5} light years (*ly*). If we use this value for **d** in Equation 1, we get $\mathbf{D} = (2.06 \times 10^{5}) * (1.5802 \times 10^{-5}\ ly)/\theta$. Multiplying the two numbers together, we get a final working equation:

$$\mathbf{D} = \frac{(3.26\ ly)}{\theta} \qquad\qquad \textit{Equation 2}$$

With this equation, we can calculate the distance to any star that is close enough to yield a detectable value of θ.

The purpose of today's exercise is to learn to use this equation to measure the distance to a number of nearby stars. We will also calculate the smallest angle that can be measured with the Hubble Space Telescope as well as the smallest angles typically measurable from the Earth's surface. Using these angles, we will ultimately calculate just how many stars there might be whose distances could be accurately measured with current technologies.

Procedure
Section 19.1

Start *RedShift College Edition*, and then bring the *RedShift 3* screen to the front. (If you've forgotten how to do this, refer to Appendix A.) From the Display menu, click **Off** everything *except* **Stars**. Click on the Filters button, and select **Stars** from the drop-down menu to open the Stars filters window. In the upper left section of the Filters window, you will find the **Magnitudes** slider. Move the upper right-hand **Magnitudes** slider to the left until it indicates a value of **1.5**. In the section below the Magnitudes slider, you will find the **Luminosity classes** slider. Move the left-hand

Luminosity classes slider to the right to **III**. Finally, make sure the **Restrict display to:** box reads **Tycho and Hipparcos**.

When the Stars filters window looks like the illustration shown here, click the **OK** button. This will display only the bright stars that are of normal size and thus are relatively nearby. Change the Zoom to **0.3**, and press the <Enter> key on the computer keyboard to show you the whole sky. Now you should see 10 to 15 stars on the screen.

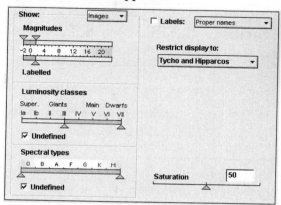

Section 19.2

One by one, center the cursor over each star, and click on it to open its name tag.   Click on the name tag to open the information box, and then click on the **More** button to expand the box. Look through the information here until you come to an entry titled **Parallax**. Record the proper name of the star along with the parallax in the data table on page 212. The units of the parallax angle look like a quotation mark: ″. This symbol stands for **arc-seconds**. Repeat the process until data are recorded for all the stars on the screen or all the empty rows in the data table are filled.

Section 19.3

Now we are ready to calculate the distance, **D**, to each star using Equation 2: $D = (3.26 \ ly)/\theta$. For each star in your data table, put the parallax angle into this equation for θ, calculate the distance of the star and record this value in the distance column in your data table. Here are the calculations for the star, Sirius, as an Example:

The parallax for Sirius is $\theta = 0.375$. Putting this into Equation 2 we get

$$D = \frac{(3.26 \ ly)}{(0.375)}$$

$$= 8.7 \ ly$$

Data Table

19

Nearby Bright Stars						
Star name	Parallax (″)	Distance (ly)		Star name	Parallax (″)	Distance (ly)
Sirius	0.375	8.7				

Sample Calculation (Show each step along with the units.)

In the space below, show the calculation described in Section 19.3 for *one* of the other 15 stars. Show each step of the calculation completely, including the units of measurement for each quantity.

D =

Q1 Do smaller parallax angles yield smaller or larger distances?

Q2 Would it be easy or difficult to measure angles like those you recorded in the data table?

Q3 A very large amateur telescope (27-inch diameter) can accurately measure 0.1 arc-second angles (which is the very best our atmosphere usually allows). With this telescope, how far away would the most distant measurable star be? (Use the steps outlined in Section 19.3 of the procedure to calculate the answer.)

$$D = (3.26 \ ly)/\theta$$
$$= (3.26 \ ly)/(\qquad)$$
$$= (\qquad)$$

Q4 The bigger a telescope is, the more sharply it can focus. The smallest parallax angle that a perfectly made telescope can detect is given by the formula

$$\theta = (2.5 \times 10^5)\lambda/2\mathbf{D}$$

212

Here λ is the wavelength of light (on average 5500 × 10⁻¹⁰ m) and **D** is the diameter of the telescope mirror. The Hubble Space Telescope has a diameter of 2 meters and is above our atmosphere. What is the smallest parallax this telescope could measure?

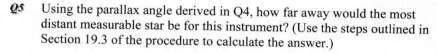

$$\theta = (2.5 \times 10^5)(5500 \times 10^{-10} \, m)/(2\mathbf{D})$$
$$= (2.5 \times 10^5)(5500 \times 10^{-10} \, m)/(2 * (\qquad))$$
$$= (\qquad)$$

Q5 Using the parallax angle derived in Q4, how far away would the most distant measurable star be for this instrument? (Use the steps outlined in Section 19.3 of the procedure to calculate the answer.)

$$D = (3.26 \, ly)/\theta$$
$$= (3.26 \, ly)/(\qquad)$$
$$= (\qquad) \, ly$$

Q6 In the region of space out to 12 light years from the Earth, there are only 30 stars. That's not very many stars for a lot of space! This works out to a density of approximately **4.1 × 10⁻³** stars per cubic light year of space. Our last problem will be to figure out approximately how many stars are close enough for the Hubble Space Telescope to accurately measure their distances.

Your answer to Q5 gives you the maximum measurable distance for the Hubble Space Telescope. To find out about how many stars are in that volume of space, we must first calculate the volume of space encompassed by that radius.

Put the distance derived in Q5 into this formula for **r**:

$$V = \frac{4}{3}\pi \mathbf{r}^3$$
$$= \frac{4}{3}\pi(\qquad)^3$$
$$= (\qquad) \, ly^3$$

The Hubble Space Telescope could accurately measure the distance to any star in this volume of space.

Q7 To get the total number of measurable stars, multiply the volume you calculated in Q6 by the average density of stars in our neighborhood, 4.1 × 10⁻³ *stars/ly³*.

19

Total number of measurable stars = **(Q6)**$*(4.1 \times 10^{-3}$ *stars/ly*$^3)$
= (*ly*$) * (4.1 \times 10^{-3}$ *stars/ly*$^3)$
= (   ) *stars*

This is approximately the total number of stars whose distances could be accurately measured by the Hubble Space Telescope.

Discussion

As you have seen, even the nearest stars are very far away. Astronomers have searched for stellar parallaxes since the days of the early Greeks, but only since the 1830's have we had instruments capable of detecting such small angular changes in position. Even today, with our finest instruments, we can accurately measure the distances to only a tiny fraction of the stars we can see. Anything farther away than the distances you calculated in today's lab must be approximated using other methods.

However, we are fortunate that in the volume of space we *can* measure, there is a wonderful variety of stars. By measuring the distances and characteristics of these relatively nearby stars and understanding certain facts about how stars produce their heat and light, we can then estimate the distances to most of the other stars in our galaxy with a fairly high degree of accuracy. In the next few exercises, you will learn about some of the other characteristics of stars that enable us to overcome some of the limitations of these great distances that separate us from the stars.

Conclusion

On a separate sheet of paper, summarize what you have learned about the distances to the stars. Begin by describing the method of parallax, and then tell the range of distances we measured in our lab today. Mention the factors that limit our ability to measure these distances.

Also include your approximation of how many stars might be measurable with current technology. Finally, suggest a couple of things we could do with our current technology to improve our abilities to measure the distances to the stars.

Star Brightness

Introduction and Purpose

The purpose of today's lab is to consider some of the characteristics of bright stars. First, we will discuss the factors that affect a star's brightness. Then we will define a standard to measure the absolute brightness of a star. Finally, we will take a closer look at the stars that are bright and discover how these factors work together to affect their brightness.

Procedure

Section 20.1

Start *RedShift College Edition*. Click the large **Go** button in the navigation bar, select **Science of Astronomy**, **The Astronomer's Tool Chest**, and **Light**, and then click on page **21**. Read and do the experiments through page **23** to discover two important phenomena that affect the brightness of a star.

Q1 Name two principal factors that affect the *apparent brightness* of a star.

Section 20.2

Click the **Go** button in the navigation bar, select **Science of Astronomy**, **Stars**, and **Types of stars**, and then click on page **4**. Read and do the experiments through page **8** to discover another important characteristic of a star that affects its brightness.

Q2 Name another factor that affects the brightness of a star.

Section 20.3

Click the **Go** button in the navigation bar, select **Science of Astronomy**, **Stars**, and **Lifestyles of the stars**, and then click on page **11**. Read, listen, and do the experiments through page **15** to discover another important characteristic of a star that affects its brightness.

Q3 Name one other factor that affects the brightness of a star.

Section 20.4

From the Window menu, select *RedShift* **3/***RedShift* **college edition** to bring the *RedShift 3* screen to the front. From the Information menu, select **Find** to open one of *RedShift's* most powerful tools, the Find window. When the Find window opens, click on the small arrow to the

20

right of the **Find:** box (which initially reads **All**), select **Stars**, and then click on **Proper names**. Now you should see a list of star names in the left-hand list box.

In the list box, find each of the stars shown in the data table below, and double-click on the star's name to select the star. When the name has been selected, click on **Information on....** in the **List Links** box to open an information box for the star. Once the information box is open, click on the **More** button to expand the box. Then select the **Color & luminosity** tab. Find the distance, spectral type (including the Roman numerals), and absolute magnitude (round off to 1 decimal place), and record this information in the data table below. The second half of the data table is done as an example. Then answer questions Q4–Q9.

Data Table

Bright Stars > 200 Light Years from Earth								
Star name	Distance (ly)	Spectral class	Absolute magnitude		Star name	Distance (ly)	Spectral class	Absolute magnitude
Acrux					Mintaka	916	O9 II	-5.0
Adhara (Adara)					Mirach	199	M0 III	-1.7
Agena					Mirphak	592	F5 I	-4.4
Almaak					Murzim	500	B1 II	-4.0
Alnilam					Naos	1400	O5 I	-6.0
Alnitak					Nunki	224	B2 V	-2.0
Alphard					Polaris	431	F7 I	-3.5
Alsuhail					Sadr	1524	F8 I	-6.0
Antares					Saiph	722	B0 I	-4.7
Avior					Sargas	272	F1 II	-2.7
Becrux					Shaula	703	B1 IV	-5.0
Bellatrix					Shedir	229	K0 II	-1.8
Betelgeuse					Spica	262	K1 V	-3.6
Canopus					Suhail Al Muhlif	841	O9 I	-5.3
Deneb					Turais	692	A8 I	-4.3
Dschubba					Wezen	1792	F8 I	-6.7

Q4   Which of the primary star types, O, B, A, F, G, K, and M, is the most common in this list?

216

Q5 Which of the three factors listed in Q1 through Q3 would explain the brightness of these stars?

Q6 Which of the primary star types, O, B, A, F, G, K, and M, are missing from this list?

Q7 Why should these stars not be easily visible so far away?

Q8 If two stars appear equally bright to us, but one is much farther away, which star is actually the brighter? Which star has the greater luminosity?

Q9 Considering only the temperature factor, which star type in this list probably should not be visible at such large distances?

Section 20.5

Click on the Filters button, and select **Stars** from the drop-down menu. In the Filters window, look at the **Luminosity classes** section, and note the titles above the various Roman numerals. Now look back at the data table and answer the following questions.

Q10 After the letter in each spectral type classification, there is a number and a Roman numeral. Which Roman numeral classification is most common in our data table?

Q11 What is the largest Roman numeral represented in our data table?

Q12 What do your answers to Q10 and Q11 tell you about these bright, distant stars' diameters?

Q13 How do your answers to Q10 and Q11 help explain why these stars are bright?

Q14 Given that the Roman numerals in the spectral class represent the diameter of the star, explain how a cool star can be just as bright as a hot one.

Q15 If two stars with different temperatures have the same luminosity, which would have the larger diameter, the hot star or the cool one?

Section 20.6

Absolute magnitude is the astronomer's way of comparing stars equally. It is a measurement of what the star's brightness *would* be if it were 10 parsecs from the Earth. Our sun, for instance, would look like a 5th magnitude star if viewed from 10 parsecs away—invisible from most urban neighborhoods!

20

In case you didn't know (or have forgotten), a difference of one magnitude corresponds to a difference of approximately 2½ times in brightness. So a 4th magnitude star would be about 2½ times brighter than a 5th magnitude star. A 3rd magnitude star would be about 2½ times brighter than a 4th magnitude one, etc. The smaller the magnitude number (including negative values), the brighter the star. Now look back at the data table, and answer the following questions.

Q16 Which of the stars in the data table is actually the brightest?

Q17 How many times brighter is this star than our Sun? To do this calculation, first subtract the star's magnitude from the Sun's (+5). For instance, Acrux, with a magnitude of −4, would be +5 − (−4) = +9 magnitudes brighter than the Sun.

Then multiply 2.5 times itself once for each magnitude. So, for Acrux, the calculation would look like this:

2.5 × 2.5 × 2.5 × 2.5 × 2.5 × 2.5 × 2.5 × 2.5 × 2.5 (9 times).

You can do this calculation much more easily if your calculator has a y^x (or x ^ y) button. Punch in **2.5**, press the y^x (or **x ^ y**) button, punch in the magnitude difference (in this case **+9**), and then press the = button. Doing this calculation for Acrux shows that it would be 3814 times brighter than our Sun if they were both at the same distance from Earth. In the space below, do this calculation for the brightest star in the data table.

Q18 Which star in the data table is actually the faintest?

Q19 How many times brighter is this star than our Sun? Do this calculation just as you did in question 17.

Discussion

The bright stars we see in the sky at night, although not average, are excellent examples of the different kinds of stars that exist and of the factors that create those differences. We find the effects of temperature, size, and distance all profoundly illustrated in the appearance of these bright stars. So far, however, we do not have enough information to describe *why* these differences exist. In the exercise "The H-R Diagram," you will discover some of the *causes* of the different temperatures and sizes of these stars.

For now, however, let's review the factors that affect a star's brightness. The *actual* brightness of a star is determined mainly by two factors: temperature and size. The relationship between the brightness of a star and its temperature is stated mathematically in the **Stefan-Boltzmann relation**. The Stefan-Boltzmann relation states that the energy output of a star is proportional to the fourth power of the star's temperature. $E \propto T^4$. So if one star is twice as hot as another (and everything else is the same), its light output will be $(2)^4$, or 16 times brighter than the other's.

Second, a star's *actual brightness* is proportional to its surface area, $4\pi r^2$ (or to the square of its diameter). So if one star is twice as large as the other (and everything else is the same), it will be $(2)^2$, or 4 times as bright.

Finally, in addition to these two factors, the *apparent brightness* of a star as seen from the Earth is also inversely proportional to the square of its distance from us. So if two stars appear to be the same brightness but one is 10 times farther away, the more distant star is actually $(10)^2$, or 100 times brighter than the closer one. In order to fairly compare the brightness of stars at different distances, astronomers have defined a measurement of a star's actual brightness called **absolute magnitude**. The absolute magnitude of a star is how bright it would appear to be if it were 10 **parsecs** from Earth. (A parsec is the distance corresponding to a parallax of one arc-second.)

Conclusion

In a few paragraphs, summarize what you have observed in today's lab about the bright stars in our night sky. Consider the following questions: What are the factors affecting the apparent brightness of a star? What is the difference between the **apparent magnitude** of a star and its **absolute magnitude**? What do most of the bright, distant stars have in common? What are the exceptions? How can you explain why these exceptional stars are also bright?

Notes

The Bright Stars and the Neighborhood

Introduction and Purpose

The purpose of today's lab is to discover what the brightest stars in the sky are like—what their colors and temperatures are and how far away they are. Along the way we will see just how many of these bright stars are relatively nearby, and finally compare these most noticeable stars to our sun's other neighbors.

Procedure

Section 21.1

Start *RedShift College Edition*, and then bring the *RedShift 3* screen to the front. From the Display menu, turn off **Natural Sky Color**, **Milky Way**, **Sun & Planets**, **Moons**, **Constellations**, **Deep Sky**, and **Guides**, so that only the stars are left turned on. Click on the Filters button, and select **Stars**. Move the upper right-hand **Magnitudes** slider to the left to **2.0**. Now click the **OK** button. This will display only the bright stars (brighter than 2nd magnitude). These are the stars that are easily visible from a typical city or suburban neighborhood.

Click on the small arrow to the right of the Aim box and select **Sun** from the menu that drops down. Click on the small arrow to the right of the Zoom box, select **0.26** from that drop-down menu, and press the <Enter> key on your keyboard. Now you should see around 50 stars on the screen.

Section 21.2

One by one, click on each star to show its name tag. Look at the star's name. If this star is already recorded in the data table on page 222, skip to the next one. If not, click on the name tag to open the star's information box. Click on the **More** button to expand the information box, and then click on the **Color & Luminosity** tab to show the star's spectral type. In the information box, find the distance to the star. If the star is less than 150 light years away, record the star's data in the data table's first column; otherwise, record it in the second column. Record the distance in the first column and the first letter and number of the spectral type in the second column. For example, the star Arcturus is listed as a spectral type **K2IIIp**, but its spectral type is recorded in the data table as **K2**. Repeat this process until all the stars shown on the screen have been identified or all the blank spaces in the Brightest Stars data table have been filled.

Brightest Stars Data Table

21

Bright Stars < 150 ly				Bright Stars > 150 ly		
Star Name	Distance	Spectral Type		Star Name	Distance	Spectral Type
Aldebaran	65	K5		Acrux	321	B0
Alhena	85	A0		Al. TrA.	415	K2
Alnair	101	B7		Antares	604	M1
Altair	17	A7		Betelgeuse	428	M2
Arcturus	37	K0		Becrux	353	B0
Cappella	42	G8		Bellatrix	243	B2
Castor	52	A2		Canopus	313	F0
Del. Vel.	80	A1		Mirphak	592	F5
Fomalhaut	25	A3		Peacock	183	B2
Gacrux	88	M4		Polaris	431	F7
Hamal	66	K2		Sargas	272	F1
Kaus Australis	145	B9		Suhail Al Muhlif	841	WC8
Sirius	9	A1		Wezen	1792	F8

Section 21.3

In the table below, you will find a list of the 32 nearest stars to the Earth. Look through it, and mentally compare the star types in this list with those in the table you just compiled. Then answer the following questions.

Nearest Stars to the Earth

Star Name	Distance (ly)	Type	Lumin. Sun = 1	Star Name	Distance (ly)	Type	Lumin. Sun = 1
Sun	0	G2	1.0	Gliese 15 A	11.3	M1	0.006
Proxima Centauri	4.2	M5	0.000006	Gliese 15 B	11.3	M3	0.0004
Alpha Centauri A	4.3	G2	1.5	Epsilon Indi	11.3	K5	0.14
Alpha Centauri B	4.3	K0	0.5	61 Cygni A	11.3	K5	0.08
Barnard's Star	6.0	M4	0.0004	61 Cygni B	11.3	K7	0.04
Wolf 359	7.8	M6	0.00002	Gliese 725 A	11.4	M3	0.003
Lalande 21185	8.2	M2	0.005	Gliese 725 B	11.4	M4	0.002
Luyten 726-8 A	8.6	M5	0.00006	Tau Ceti	11.4	G8	0.45
Luyten 726-8 B	8.6	M6	0.00004	Procyon A	11.4	F5I	7.7
Sirius A	8.6	A1	24	Procyon B	11.4	wd	0.0006
Sirius B	8.6	wd	0.003	Lacaille 9352	11.5	M2	0.01
Ross 154	9.6	M4	0.0005	GJ 1111	11.8	M7	0.00001
Ross 248	10.3	M6	0.0001	GJ 1061	12.0	M5	0.00008
Epsilon Eridani	10.7	K2	0.3	Luyten 72532	12.2	M5	0.0003
Ross 128	10.8	M4	0.0003	Gliese 273	12.3	M4	0.001
Luyten 7896	11.1	M5	0.0001	Gliese 825	12.5	M0	0.03

Q1   Compare the nearest stars to the Earth with the Brightest Stars data table. How many of the 32 closest stars are also among the brightest?

Q2   Which spectral type is the most common in the **Nearest Stars** data table?

Q3   Which spectral type is the most common in the **Brightest Stars** data table in stars closer than 150 light years?

Q4   Which spectral type is the most common in the **Brightest Stars** data table in stars more distant than 150 light years?

Q5   How many stars more distant than 1000 light years are bright enough to be seen easily from the Earth?

Section 21.4

In the following table, you will find a review of the information you learned in the last exercise comparing the temperatures and colors of the various spectral types. Look through it, and mentally compare the

21

temperatures (and colors) of the brightest stars and then of the nearest ones. Then answer the following questions.

Spectral Types, Temperatures, and Colors							
Temperature (K)	>25,000	11,000-25,000	7,500-11,000	6,000-7,500	5,000-6,000	3,500-5,000	<3,500
Spectral type	O	B	A	F	G	K	M
Color	Blue	Blue-white	White	Yellow-white	Yellow	Orange	Red

Q6 How many extremely hot stars (type O) are in your data table that are less than 1000 light years away?

Q7 How many very hot stars (type B) are in your data table that are less than 1000 light years away?

Q8 How many O or B stars are in the Nearest Stars data table?

Q9 How many fairly hot stars (type A) are in your data table that are less than 1000 light years away?

Section 21.5

Now, let's see if the star composition far away from our Sun is similar to that of our local neighborhood. We will use our close neighborhood stars to predict what the star composition *should* be way out at 1000 light years. Then we will see if our predictions hold true. We will start by calculating how many G stars there are for every A star in our neighborhood.

Q10 In the Nearest Stars table, how many A stars are there?

Q11 In the Nearest Stars table, how many G stars are there?

Q12 In the Nearest Stars table, what is the ratio of G stars to A stars? (Divide the number of G stars by the number of A stars.)

$$\frac{Q11}{Q10} =$$

$$\frac{(\qquad)}{(\qquad)} =$$

Q13 If the region of space out to 1000 light years is similar in makeup to that of the nearby stars, it should have the same ratio of G to A stars as you calculated in Q12. If this is so, you should be able to predict how many G stars should be out there.

A little algebra (as shown below) gives the number of G stars within 1000 ly. Just multiply the ratio of G to A stars (which you calculated in Q12) by the number of A stars in the surrounding 1000 ly region of space (which you measured in Q9).

$$\frac{|G\ stars\ |}{|A\ stars\ |} * A\ stars_{1000ly} = G\ stars_{1000ly}$$

Q12 * Q9 = G stars $_{1000ly}$

(   ) * (   ) =

Section 21.6

Now let's do the same calculation for K stars.

Q14 In the Nearest Stars table, what is the ratio of K stars to A stars?

$$\frac{\textbf{K\ stars}}{\textbf{Q10}} =$$

$$\frac{(\qquad)}{(\qquad)} =$$

Q15 If the region of space out to 1000 light years is similar in makeup to that of the nearby stars, predict how many K stars should be out there. As with the G stars, multiply the ratio of K stars to A stars (Q14) by the number of A stars within 1000 ly (Q9).

Q14 * Q9 =

(   ) * (   ) =

Section 21.7

Now let's repeat the calculation one more time for M stars.

Q16 In the Nearest Stars table, what is the ratio of M stars to A stars?

$$\frac{\textbf{M\ stars}}{\textbf{A\ stars}} =$$

$$\frac{(\qquad)}{(\qquad)} =$$

Q17 If the region of space out to 1000 light years is similar in makeup to the nearby stars, predict how many M stars should be out there. As before, multiply the ratio of M stars to A stars (Q16) by the number of A stars within 1000 ly (Q9).

Q16 * Q9 =

(   ) * (   ) =

Section 21.8

Now we reach the moment of truth. Let's compare the predictions we just made (based on the nearest stars) with the actual numbers of each type of star recorded in the Brightest Stars data table.

21

Q18 Compare the Nearest Stars predictions (Q13, Q15 and Q17) with the actual counts for G, K, and M stars from the Brightest Stars data table.

	Predicted by Nearest Stars' Averages	Actual Numbers From Brightest Stars Data Table
G stars	(Q13)	
K stars	(Q15)	
M stars	(Q17)	

Q19 Our predictions weren't very accurate, were they? This could mean that the local neighborhood is quite different from the rest of space, or there could be some other reason for this big discrepancy.

We don't see any other evidence to suggest that our neighborhood should be different from the rest of the galaxy, so let's look for another possibility. What are the typical luminosities of the G, K, and M stars in the Nearest Stars data table?

Q20 Compare the luminosities you found in the last question with those in the Brightest Stars data table. If there were lots of these G, K, and M stars far away, would you be able to see them?

As you have seen, there are many cool, dim stars in the space right around our Sun. There could easily be many of these stars all throughout the galaxy, and we would never see them. As you will see in future exercises, however, they do show up in photographs taken through large telescopes.

Discussion

Our Sun is a type G2 star. In a radius of 26 light years, there are only 3 A-type stars, one F-type star, one other G-type star, and numerous K- and M-type stars. Yet the truly brightest stars are the really hot ones—O's, B's, and A's. Apparently, really big, hot stars are the rarity rather than the exception in our part of the galaxy, yet these are the stars we notice the most in the sky!

Looking at the table of the nearest stars, however, you will notice lots of cool stars. Look at their luminosities. The reason these are not in the brightest list is that they are intrinsically small and therefore both cool and dim. If the rest of our neighborhood is like this small sample, then there must be hundreds of these small, dim, cool stars for every large, noticeable star in our brightest list.

Conclusion

In a few paragraphs, summarize what you have observed in today's lab about the brightest stars in our night sky. In your summary, consider the following questions: What do most of the bright stars have in common? What are the exceptions?

Are these bright stars average stars or extraordinary ones relative to the stars nearest the Solar System. Given the information presented here, what can you say about the relative likelihood in our region of the galaxy of forming large stars, forming medium-sized ones, and forming small ones?

Notes

21

Name: _____

Course: _____

Section: _____

The H-R Diagram

Introduction and Purpose

In the last few exercises, we have examined at length both the nearby stars and the bright stars. We have seen that these stars come in many different sizes and temperatures and are found at many different distances from the Earth. In the exercise "Star Brightness," we discovered that a star's luminosity depends primarily on two characteristics: size and temperature.

In today's exercise, we will attempt to explain the relationship between these two factors. We will also introduce a scheme for graphically plotting the characteristics of brightness and temperature called the Hertzsprung-Russell diagram (H-R diagram for short). We will then look for relationships (or patterns) between these two factors.

Procedure
Section 22.1

Start *RedShift College Edition*. Before we proceed, we need to look at how a star produces its heat. Click the **Go** button in the navigation bar, select **Science of Astronomy**, **Stars**, **What Is a Star?**, and click on page **8**. Read, watch, and do the experiments through page **29** to discover how a star produces its light and heat.

In a normal star, there are two processes at work against each other—gravity and nuclear fusion. Gravity tries to crush the enormous mass of the star into an incredibly dense core. As gravity crushes the core to higher and higher densities, the atoms in the core collide more and more violently with each other, creating higher and higher temperatures and pressures. When the temperatures in the core reach 10 million Kelvin, head-on collisions between hydrogen atoms are so violent that the nuclei merge, resulting in tremendous explosions; this is called nuclear fusion. At even higher temperatures, more massive atoms like helium, carbon, and oxygen can fuse, producing even more violent explosions.

So gravity, on the one hand, tries to crush the star out of existence. At the same time, however, this crushing force of gravity triggers a second force: nuclear fusion. The nuclear fusion, then, tries to blow the mass of the star back out into space, thus partially counteracting the force of gravity. When a star first forms, these two forces fluctuate until they finally become balanced. Once they are balanced, the star will "fuse" hydrogen in a stable manner until one or the other of the forces changes significantly.

22

Section 22.2

In the data table below, you will find two shortened lists we have already encountered—a list of the nearest stars to the Earth and a list of the brightest stars as seen from the Earth. Once you have familiarized yourself with the data table, plot the luminosity of each star versus its temperature on the graph on the next page. All the star entries that are shaded in the data table have already been plotted for you as examples and are indicated on the graph with superscripts.

Data Table

Nearest Stars to the Earth				Brightest Stars in the Sky			
Star Name	Distance (ly)	Spectral Type	Lumin. Sun = 1	Star Name	Distance (ly)	Spectral Type	Lumin. Sun = 1
Sun	0	G2V	1.0	Canopus [8]	98	F01	1×10^3
Proxima Centauri	4.2	M5V	6×10^{-6}	Arcturus	36	K2III	1×10^2
Alpha Centauri A [1]	4.3	G2V	1.5	Vega [9]	26	A0V	5×10^1
Alpha Centauri B	4.3	K0V	5×10^{-1}	Capella	46	G8III	2×10^2
Barnard's Star [2]	6.0	M4V	4×10^{-4}	Rigel [10]	815	B8I	8×10^4
Wolf 359	7.8	M6V	2×10^{-5}	Procyon	11	F5IV	9
Lalande 21185 [3]	8.2	M2V	5×10^{-3}	Betelgeuse [11]	425	M2I	1×10^5
Luyten 726-8 A	8.6	M5V	6×10^{-5}	Achernar	65	B3V	5×10^2
Luyten 726-8 B [4]	8.6	M6V	4×10^{-5}	Beta Centauri [12]	300	B1III	9×10^3
Sirius A	8.6	A1V	2×10^1	Altair	17	A7V	1×10^1
Ross 154 [5]	9.6	M4V	5×10^{-4}	Aldebaran [13]	20	K5III	2×10^2
Ross 248	10.3	M6V	1×10^{-4}	Spica	260	B1V	6×10^3
Epsilon Eridani [6]	10.7	K2V	3×10^{-1}	Antares [14]	390	M1I	1×10^4
Ross 128	10.8	M4V	3×10^{-4}	Pollux	39	K0III	6×10^1
Luyten 7896 [7]	11.1	M5V	1×10^{-4}	Fomalhaut [15]	23	A3V	5×10^1
Gliese 15 A	11.3	M1V	6×10^{-3}	Deneb	1400	A2I	8×10^4

Note: Plot only those stars in the data table that are *not* shaded. The shaded ones are plotted for you as examples of how to plot different luminosities.

Graph

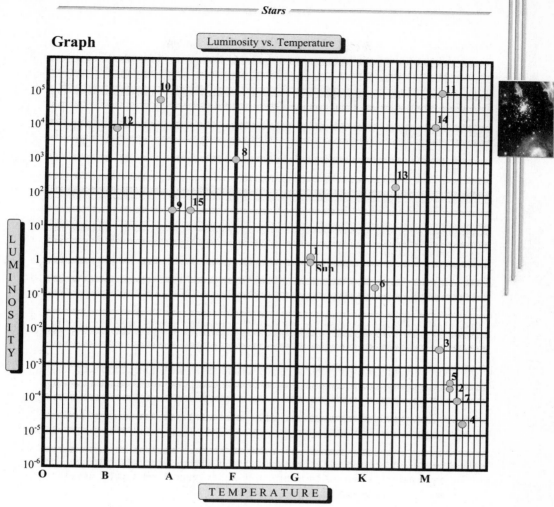

Luminosity vs. Temperature

To plot each star, first find the vertical line that represents the temperature of the star. For example, the Sun's **G2V** is plotted 2 lines to the right of G. Then, starting in the center of that line at the **1**, go either up or down depending on whether the star's luminosity is larger than the Sun's or smaller. For example, Vega's (star **9**) luminosity is 5×10^1, which is halfway between 10^1 and 10^2. Now look at star **2** on the graph. Its luminosity has a negative power of 10 (less than that of the sun)—4×10^{-4}. This value is plotted almost halfway up between the 1×10^{-4} and 1×10^{-3} lines.

Section 22.3

Click on the Go button in the navigation panel, select **Science of Astronomy**, **Stars**, and **Lifestyles of the stars**. Click on page **1**. Read, watch, and do the experiments through page **10** to discover what the H-R diagram tells us. Then answer the following questions.

22

Q1 Based on what you just read, determine which should have a higher rate of nuclear reactions at its core, a more massive star or a less massive one. (In which one would gravity crush the core more?)

Q2 Given your answer to the previous question, should a very massive star be very hot or very cool?

Q3 Given your answer to Q1, should a star with a very small mass be very hot or very cool?

Q4 If a star has a very large mass, would it probably have a large diameter or a small diameter?

Q5 If a star has a very small mass, would it probably have a large diameter or a small diameter?

Q6 Given your answers to Q2 and Q4, should a star with a large mass be bright or dim?

Q7 Given your answers to Q3 and Q5, should a star with a small mass be bright or dim?

Q8 Given your answers to Q2 and Q6, in what part of the graph would an extremely massive star most probably lie? (Consider what its brightness and temperature would be.)

Q9 Given your answers to Q3 and Q7, in what part of the graph would a star with a very small mass most probably lie? (Consider what its brightness and temperature would be.)

Q10 Given your answers to Q8 and Q9, if you plotted a lot of stars on the graph with a whole range of different masses, what would the graph look like?

Q11 You should see a curving, diagonal pattern in the stars plotted on your H-R diagram. But there are several stars that do not lie along this path. In what part of the graph are most of these stars?

Q12 Which star on your graph is hottest?

Q13 Which star on your graph is dimmest?

Q14 Which star on your graph is both very cool and extremely bright?

Q15 Review your answers to Q8 and Q9. From its position on the H-R diagram, would you expect Vega to be more or less massive than our Sun?

Q16 Review your answers to Q8 and Q9. From its position on the H-R diagram, would you expect Epsilon Eridani to be more or less massive than our Sun?

Q17 On your graph, indicate the main sequence by drawing a line all the way around those stars that are clearly on the main sequence.

Q18 Are more of the stars plotted on your graph on the main sequence or off the main sequence?

Section 22.4

Again, click on the **Go** button in the navigation panel, and select **Science of Astronomy**, **Stars**, and **Lifestyles of the stars**. Click on page **11**. Read, watch, and do the experiments through page **16** to learn about the stars in the upper right-hand part of the H-R diagram.

Q19 What term is used to describe those stars on your graph that are not main sequence stars?

Q20 Roman-numeral suffixes of I, II, or III in the Spectral Type column denote giant stars. Do the Roman-numeral suffixes of the non-main-sequence stars on your graph indicate that those stars are giants?

Q21 Review the normal star processes discussed in Section 22.1. If these are the only explanation for a star's temperature and brightness, which stars would seem to be brighter than they should be—the main-sequence stars or the non-main-sequence stars?

22

Discussion

Notice how the stars on your graph seem to lie on a few well-defined lines. The central line curving from the upper left diagonally down to the lower right is called the ***main sequence***. Semi-horizontal lines branching off the main sequence to the right are called the giant branches. These lines were discovered when Hertzsprung and Russell set out to plot the actual brightness of various stars (their luminosities) relative to their colors (or temperatures). Every time they plotted a new star's brightness versus its color, the star ended up lying on either the main sequence line, one of the giant branch lines, or in the white dwarf region. The more stars they plotted, the more apparent this linear pattern became. Most of the stars they plotted fell somewhere along the diagonal line that has come to be known as the ***main sequence***.

Most of the really bright stars, however, do not lie on the main sequence but congregate along one or another of the giant branches splitting off to the right of the main sequence. Most of these stars are cooler and therefore do not put off as much energy per square meter of surface as the hotter stars. But these stars are so incredibly huge that the decrease in light output per square meter is more than made up for by the enormous surface area of the star.

The main sequence positions on the H-R diagram are readily explained if you understand the nuclear processes we discussed in Section 22.1. The giant branches, however, clearly do not fit our explanation. If a star must be very massive to be large, why isn't it also very hot? We will answer this question in an upcoming exercise.

Conclusion

On a separate sheet of paper, write a short summary describing what you have learned about the H-R diagram. In your summary consider the following questions. Why are the coolest stars mostly very dim? Why are the hottest stars mostly very bright? Where are the majority of the stars—on the main sequence or off? Where are most of the bright stars—on the main sequence or off? How does the H-R diagram help us distinguish between ordinary stars using ordinary nuclear processes and those that are unusual?

Name: _____

Course: _____

Section: _____

Old Age and Red Giants

Introduction and Purpose

In the previous exercise, you learned that there is a pattern to the relationship between the brightness of stars and their temperatures. This relationship was first noticed by Hertzsprung and Russell, and the chart developed to graphically display it is called the ***H-R diagram***. You also learned that most of the stars we can see lie along the central diagonal line called the ***main sequence***. Finally, you learned that the reason for the main sequence was the basic process of nuclear fusion of hydrogen that is driven by the enormous gravity of the star.

Many stars, however, do not lie on the main sequence. Their brightness and temperature cannot be explained *only* by the nuclear fusion of hydrogen. The purpose of this exercise is to discover how these non-main-sequence stars relate to those on the main sequence and to see how astronomers use the H-R diagram to determine what stage of life any particular star is in.

In today's exercise we will try to identify the branches of the H-R diagram. We will discover which bright stars in the sky belong to which branch, and then see if we can use our knowledge of a star's life sequence to estimate how close to the end of its life a particular star is.

Procedure
Section 23.1

Start *RedShift College Edition*. We will begin with a recap of the process of nuclear fusion that powers every star during most of its lifetime. Click on the **Go** button in the navigation panel, select **Science of Astronomy**, **Stars**, and **What is a star?**, and click on page **8**. Read through page **19** for a review of how stars produce their energy.

Now skip to page **22**, and read through page **29** to discover why stars don't blow themselves apart. Remember that a star exists because of a balance between gravity that is trying to crush the star into a tiny ball, and nuclear fusion that is trying to blow the star apart. So long as the two forces are balanced, the star continues to destroy hydrogen and create helium. Stars like this are main-sequence stars.

23

Section 23.2

Before you proceed to the data table, you need to a quick review of the H-R diagram. Click on the **Go** button in the navigation panel, select **Science of Astronomy**, **Stars**, and **Lifestyles of the stars**, and click on page **2**. Read through page **3** for a review of the H-R diagram and the main sequence.

Now look at the data table below. Plot the luminosities of each of the stars in the data table versus their temperature types as you did in the exercise "The H-R Diagram." (As before, the Sun and two other stars are plotted for you as examples.)

Data Table

Brightest Stars in the Sky							
Star Name	Spectral Type	Lumin. Sun = 1	Type of Star	Star Name	Spectral Type	Lumin. Sun = 1	Type of Star
Sun	G2 V	1	Main sequence	Betelgeuse	M2	1×10^5	
Sirius A	A1 V	2×10^1	Main sequence	Agena	B1 III	9×10^3	Blue Giant
Canopus	F0	1×10^3		Altair	A7	1×10^1	
Arcturus	K2	1×10^2		Aldebaran	K5	2×10^2	
Rigel Kentaurus	G2 V	1.5	Main sequence	Spica	B1	6×10^3	
Vega	A0	5×10^1		Antares	M1	1×10^4	
Capella	G8 III	2×10^2	Yellow Giant	Fomalhaut	A3	5×10^1	
Rigel	B8	8×10^4		Pollux	K0	6×10^1	
Procyon	F5	9		Deneb	A2	8×10^4	
Achernar	B3	5×10^2					

H-R Diagram

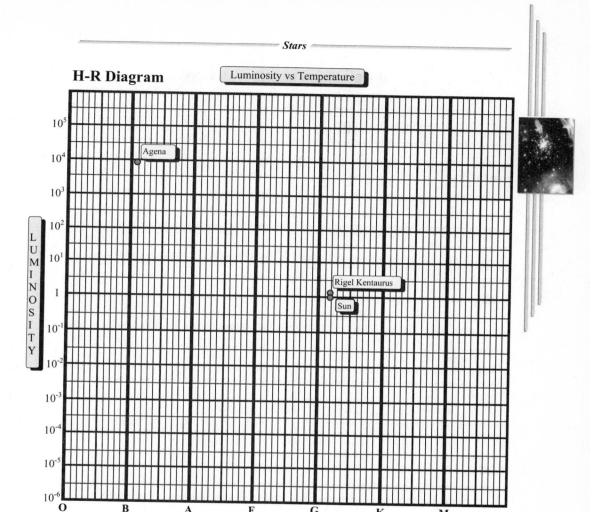

Luminosity vs Temperature

Agena

Rigel Kentaurus

Sun

LUMINOSITY

10^5
10^4
10^3
10^2
10^1
1
10^{-1}
10^{-2}
10^{-3}
10^{-4}
10^{-5}
10^{-6}

O   B   A   F   G   K   M

TEMPERATURE

Section 23.3

Once you have plotted all the stars, draw a line around the main sequence stars and label the area as "Main sequence." Then write **Main Sequence** in the Type of Star column for each of these stars. For all the other stars in the table, you now need to determine what kinds of stars these are.

To do this, bring the *RedShift 3* window to the front. From the Information menu, select **Find** to open the Find window. In the **Find:** box, select **Stars**, and click on **Proper Names**. Now select each of the non-

23

main-sequence stars in the table by scrolling down to the name and then double-clicking on it. When each star is selected, click **Information on…** in the List Links box, and click **OK** to open the star's information box. Click on the **More** button to expand the information box, and then select the **Color & Luminosity** tab.

Look at the Spectral type entry, and note the Roman numeral suffix. (Some stars have two groups of letters following the spectral type. Use the first group of Roman numerals, ignoring suffixes such as "a" or "b.") Write this Roman numeral in your data table in the empty space in the Spectral Type column. Note the color of the star, and jot that down in a notepad for use in the next section. Finally, close the star's information box. Repeat this procedure until all the non-main-sequence stars have been identified.

Section 23.4

Once all the stars have Roman numeral suffixes, click on the Filters button and select **Stars** to open the Stars filters window. In the Spectral type area of the window, note the type of star that corresponds to each of the Roman numerals.

Now go back to your data table and finish filling in the Type of Star column. For each non-main-sequence star, look at its temperature type and its Roman numeral suffix, and then, referring to your notes, write in the color and kind of star each one is. (Agena is done as an example: It is a **B** star, which is **blue**, and a **type III**, which is a giant.)

Section 23.5

Q1 Circle the linear region that represents the type I (Super-giant) evolutionary track, and label it "Super-giant." Which stars are type I super-giants?

Q2 Circle the linear region that represents the type II (Bright Giant) evolutionary track, and label it "Bright Giant." Which stars are type II giants?

Q3 Circle the linear region that represents the type III (Giant) evolutionary track, and label it. Which stars are type III giants?

Q4 As you see, many of the bright stars are giants and super-giants. However, just because a star is a giant does not mean it is not a main-sequence star. The diagonal main-sequence region of the H-R diagram goes all the way to the upper left-hand corner. Many of the giant stars, however, are clearly not on the main sequence. In the space below, list all the stars on the diagram that are significantly off the main sequence.

Section 23.6

Generically, the stars you listed in Q4 are called red giant stars. What makes these stars so different? To answer this question, click on the **Go** button in the navigation panel, select **Science of Astronomy**, **Stars**, and **Lifestyles of the stars**, and click on page **16**. Read through page **20** for a review of how a star becomes a red giant.

Eventually, every star will use up all the hydrogen in its core. As it does so, several things begin to happen. First, as the core hydrogen is being depleted, it is being replaced by the helium by-product of the hydrogen fusion. Since the temperature and pressure at the core is not at first sufficient to fuse helium, the helium core gets more and more massive. As its mass increases with no nuclear reactions to force it outward, gravity forces the core to shrink.

As the core shrinks, the helium atoms collide more and more frequently with greater force, causing the temperature of the core to shoot upward. All this heat from the center moves outward into the hydrogen around the core, forcing the hydrogen outward and causing the hydrogen fusion to slowly migrate to the outer regions of the star.

At this point, the star is becoming two stars in one. The core has now collapsed under its own weight and has become both very small and very hot. The center is turning into the first stages of a white dwarf star. At the same time, the increased heating from the center is driving the outer edges of the star farther outward, allowing them to begin to cool off. The hydrogen fusion that *was* producing most of the heat and light of the star is decreasing as the hydrogen is becoming more and more depleted. So the outside of the star is expanding and cooling off, becoming first yellow, then orange, and finally red.

This produces a strange effect on the star's brightness. On the one hand, the surface of the star is cooling (so each part of the star gives off less light); on the other hand, it is increasing in size and area (thus increasing the amount of light-producing surface). The result? The overall brightness remains approximately the same. Depending on the initial size and

23

temperature of the star, the resulting yellow, orange, and red stars may be ordinary giants (type III), bright giants (type II), or super-giants (type I). Except for the very smallest stars, however, no matter where on the main sequence a star began, as it runs out of hydrogen, it moves more or less horizontally to the right, its surface temperature growing less and less while its brightness remains more or less constant.

Section 23.7

Now, let's take a look at how a star changes as it grows older. Click on the **Go** button in the navigation panel, select **Science of Astronomy, Stars**, and **Birth and death of stars**, and click on page **20**. Read through page **29** to discover how the position of a star on the H-R diagram can tell us about its age. When you are finished, refer to the data table to answer the following questions.

Q5   Tracing the evolutionary track backwards, which of the stars in our data table were at some time (or still are) blue in color?

Q6   Which of the stars in the data table is probably nearest the end of its life?

Q7   Why did you choose the star you chose in the last question?

Q8   Which other star (or stars) in the data table may someday look very similar to Betelgeuse?

Q9   Which other star (or stars) in the data table may someday look very similar to Antares?

Q10   In a red-giant star, the core is hotter than when it was on the main sequence, but its surface is cooler. How can you explain this seeming paradox?

Q11   Which star or stars in the data table do you think the Sun will be most like when it becomes a red giant?

Section 23.8

Now insert the Extras CD in the computer's CD-ROM drive. When the *RedShift College Edition* startup screen appears, click **Start**. Then, when the Extras title screen opens, click on **Story of the Universe** and choose the section titled **Lives of the Stars**. Watch this narrated animation up to the part about planetary nebulae. Then answer the following questions.

Q12 Why do the outer layers of a star "blow away" during the last stages of its red-giant phase?

Q13 What single characteristic of a star determines its size, temperature, and color?

Q14 On average, how many times larger than our Sun are the largest stars?

Q15 On average, how much smaller than our Sun are the smallest stars?

Q16 How large in diameter will our Sun get when it becomes a red giant?

Discussion

In today's lab, we saw that many of the really bright stars in the sky are bright because they are truly giant stars. We also saw that if a star is born a giant, it probably will continue to be bright until it runs completely out of fusible elements. As long as a star is on the main sequence, it is primarily burning hydrogen in its core, and its brightness and color are determined principally by the mass of the star (as we saw in the exercise "The H-R Diagram").

However, once a star's core begins to contract and heat up, the outer hydrogen layers are driven away from the core by the internal heat, the star gets larger and the surface begins to cool off. Thus, when a star moves off the main sequence, its brightness remains fairly constant, but its color moves toward the red end of the spectrum.

What happens next is a story for the following two exercises. In "Puff-Balls and Smoke Rings," we will examine the final stages of small to average stars (less than 10 solar masses). In "Remains of the Day," we will take a look at the dramatic demise of the true giant stars that start out with more than 10 times the mass of the Sun.

Conclusion

In a few paragraphs, describe what you have learned about the path a star takes as it moves off the main sequence. Describe how you can tell if a star is still burning hydrogen in its core or if its core has begun to fuse other elements. Describe how you can look at a non-main-sequence star and guess what kind of star it was before it left the main sequence. Also tell how you can predict what kind of a star a main-sequence star will become when it runs out of hydrogen. Finally, tell why a star does not change much in brightness as it runs out of fuel.

23

Notes

Name: _____

Course: _____

Section: _____

Puff-Balls and Smoke Rings

Introduction and Purpose

In the past few exercises, we have been looking at the life of a star. We have seen how its original mass determines its basic temperature and brightness (its position on the H-R diagram) for most of its main-sequence "adult" life. We also saw how a star will move off the main sequence as it runs out of hydrogen and eventually become a red giant.

However, as we saw in the exercise "The Bright Stars and the Neighborhood," most of the stars in our galactic neighborhood are not very large. If the rest of the galaxy is like our neighborhood, very large stars must be in the minority. In our exercise today, we will explore the fate of the "silent majority"—the small, ordinary stars like our Sun that are not large enough to go through many different fuels after their hydrogen is depleted.

The purpose of today's exercise is to examine the last stages of the life of a small to average-size star like our Sun. We will see what shapes, colors, and symmetries characterize the remnants of red giant stars and try to draw some conclusions about what the star was like before it arrived at this point.

Procedure
Section 24.1

Start *RedShift College Edition*. For an overview of today's topic, click on the **Go** button in the navigation bar, select **Science of Astronomy**, **Stars**, and **Birth and death of stars**, and click on page **30**. Read and watch through page **34** to discover what will happen to our Sun when it runs out of fuel.

As we noted in the exercise "Old Age and Red Giants," every star eventually uses up all the hydrogen in its core. Throughout its life, the star has been making helium from its hydrogen, and this helium has been accumulating in the star's core. In a small to average star (less than about 10 solar masses), the gravitational crush at the core is not sufficient to fuse helium during its main-sequence life, so the helium core gets more and more massive. As its mass increases with no nuclear reactions to resist, gravity forces the core to shrink.

As the core shrinks, the helium atoms collide more frequently with greater and greater force causing the temperature of the core to go up. All this

24

heat from the center moves outward into the hydrogen around the core, forcing it outward to create the red giant star discussed in the last exercise.

Finally, the core temperature reaches around 100 million degrees Kelvin, and the helium begins to fuse. The core flashes outward in an explosion called the ***helium flash***. In an average star, the gravity holding the outer hydrogen atmosphere is weak because of the star's modest mass and its large, red giant radius. The helium flash blows the outer layer of the star's atmosphere into space, forming a shell of gas that will expand forever.

In a medium-size star where there is more helium, the star may go through several of these "flashes" as small amounts of subsequent nuclear fuels are consumed, and so several shells may be expelled from the outer atmosphere of the star before it finally runs out of fusible elements. Eventually all of these stars will run out of fuel. When this happens in an average star, the remaining core settles down to a final shrinking and cooling phase. These small, initially hot stars are called white dwarfs.

Section 24.2

Click on the **Science of Astronomy** tab and select **Photos** from the drop-down menu. When the photo gallery opens, click on the **Contents** button in the navigation bar to open the Contents frame. In the contents frame, click on **The Galaxy**, and select **Planetary nebulae** to put a list of planetary nebula photos in the main Photo Gallery window.

Scroll down to the image titled **[1115] Ring Nebula in Lyra**. Click on the small icon on the picture to enlarge it. This is one of the most famous and easily observed planetary nebulae in the sky. Observe this photo carefully, and answer the following questions.

Q1   What is the general overall shape of the nebula?

Q2   What color is most of the gas making up the nebula?

Q3   What kind of gas would this likely be? (*Hint:* What kind of gas would be at the outer edges of a star's atmosphere?)

Q4   Can you find the star that created the nebula? What color is it?

Q5   What kind of a star is this central star on its way to becoming?

Q6   How many "flashes" would you guess this star went through before settling down to its final form? (How many distinct shells of gas do you observe?)

Section 24.3

Click on the Thumbnail icon to go back to the contents of the Planetary nebula photo gallery. Scroll to the image titled **[1129] Planetary nebula in the constellation Norma**. Click on the small icon on the picture to enlarge it. Then answer the following questions.

Q7   Is this nebula in the constellation Norma [1129] spherical?

Q8   Is this nebula symmetric? If so, describe its symmetry.

Q9   What could you guess about the direction (or directions) of the "flash" that blew away the outer atmosphere of this star?

Q10 Does the expelled material in this nebula appear to be hydrogen?

Q11 How can you tell?

Section 24.4

Return to the contents of the Planetary nebula photo gallery. Scroll to the image titled **[1120] Dumbbell Nebula in the constellation, Vulpecula**. Enlarge the image, and then answer the following questions.

Q12 What features do you see in the Dumbbell Nebula that are similar to those in the Ring Nebula?

Q13 What features do you see in the Dumbbell Nebula that are similar to those in the planetary nebula in Norma?

Q14 Can you see the central star that remained after it blew away its outer atmosphere?

Q15 On the basis of its colors, would you guess the material blown away was mostly hydrogen or a mixture of several gases? Why?

Section 24.5

Return to the contents of the Planetary nebula photo gallery. Scroll to the image titled **[0188] Planetary nebula, the Helix, NGC 7293**. Enlarge the image, and answer the following questions.

Q16 This is one of the most beautiful planetary nebulae in the sky. Do you think that this nebula is closer than, farther than, or at about the same distance from us as the other nebula we have looked at? How can you tell?

24

Q17   What does the prominent gas in the material blown away by this star appear to be?

Q18   How many shells of gas do you see in this photo?

Q19   Do these shells appear to have been expelled one after another, or both at the same time? (Is one larger than the other?)

Q20   Think about the shapes of the nebulae in Norma and the Dumbbell Nebula. Try to explain why the Helix Nebula has two equal rings that are offset in space.

Q21   What color is the central star in this nebula?

Q22   What temperature would it be?

Section 24.6

Return to the contents of the Planetary nebula photo gallery. Scroll to the image titled **[3070] A complex planetary nebula, NGC 6543**. Enlarge this image and then answer the following questions.

Q23   The colors in this photo are artificially enhanced, but the color of the central star is fairly accurate. In a large telescope (17″ or larger), it appears to be a brilliant, robin's-egg blue. It is called the Cat's Eye Nebula. How many different explosions may have occurred in this nebula? (How many distinct, symmetric shells of gas can you count surrounding this star?)

Q24   How many different axes do the explosions seem to have centered on?

Q25   Return to the contents of the Planetary nebula photo gallery. Read the description of this image, and then give a probable explanation for why the nebula has the different axes that it does.

Section 24.7

Now bring the *RedShift 3* screen to the front. From the File menu, open your personal settings file. (See Appendix B if you have forgotten how.) Set the **Local date** and **time** to **9:00 pm**, **tonight**, and press the <Enter> key on the keyboard. Click on the Filters button, and select **Constellations**. Click in the **Boundaries** box to turn on the constellation boundaries, and then remove the check mark from the **Patterns** box to turn the patterns off. Now click on the **Deep Sky** tab. In the Deep Sky filters window, pull the upper right-hand **Magnitude** slider to **12** so that the fainter nebulae in today's lab will show up, and then click **OK**.

From the Information menu choose **Find** to open the Find window. Click on the small arrow to the right of the **Find:** box, select **Deep Sky**, and click on **Proper names**. Now scroll down through the list, double-click on **Dumbbell Nebula**, and then click **OK** to center the screen on the Dumbbell nebula. The Dumbbell nebula is quite small, so you will need to increase the Zoom to **50** to see it. It will look like a small, fuzzy hourglass in the center of the screen.

Now click on the Dumbbell nebula to open its name tag. Then click on the name tag to open its information box. In the information box, click on the **Report** button to open the Report window. In the report window, find the times of rising and setting tonight from your viewing location. If the object is not visible from your location at all on this date, you will see the phrase **below horizon**. Now record this data in the data table below. Decrease the Zoom to **1** to see what constellation this nebula is in, and record this as well. Repeat the above steps to find and record the data for each of the remaining planetary nebulae in the data table below.

Q26 Which of these nebulae would be easily visible *from your location* at 9:00 p.m. tonight?

Data Table

Planetary Nebulae							
Object	Constellations	Rise	Set	Object	Constellations	Rise	Set
Dumbbell Nebula				Ring Nebula			
Owl Nebula				Saturn Nebula			

Web Intrigues

(If you do not have Internet access, you may skip this section.)

Log on to the Internet and open your web browser. Type in the following URL: *http://hubblesite.org/gallery/showcase/text.shtml* to take you to the Hubble Space Telescope Showcase Gallery page, and click on the **Nebulae** link. On this page, you will find the best images of six amazing planetary nebulae—the Ring, the Cat's Eye, the Egg, the Hourglass, the Eskimo, and the

Spirograph nebulae. Browse through these images as long as you like. The variety of structures presented in these images is truly mind-boggling!

Now select one of these nebulae to research in depth. On that nebula's page, click the **Learn More** button to read a little about the nebula. Once you have read the caption, click on the **High Res Images** icon to open a page with more detail. If the page shows another small picture with a caption, find the link at the left titled **Press Release Images**, and click on it to open the images page; otherwise, you are already on the images page. Now click on the first **JPEG** icon to view a large version of the image.

Once you have had a good look at the large image, click the **Back** button to return to the images page, and click on the **Caption** icon to read about the nebula. In the space below (or with your conclusion), tell which planetary nebula you investigated and describe what scientists have learned from their study of this nebula. Be sure to mention any mysteries this image has presented to astronomers.

Discussion

We have seen that when a small to medium-size star runs out of hydrogen in its core, its outer atmosphere gradually expands while its core contracts. The core finally blows away the outermost atmosphere of the star, forming what we call a *planetary nebula*.

Overall, these planetary nebulae are more or less spherical in shape, but many also have a bipolar symmetry. In some, the explosions that blew away the gas shells appear to have occurred at different times and with different intensities in different directions. A few planetary nebulae are very complex, suggesting not only several explosions, but perhaps involving several individual stars as well.

Conclusion

On a separate sheet of paper, summarize the different forms a planetary nebula can take. Mention specific planetary nebulae as examples of these different forms. Explain why the shells of gas and dust are moving away from the star. Describe the colors you will most often encounter in the nebulae, and tell what these colors are produced by. Finally, describe the star you will find at the center of a planetary nebula, tell what color it will usually be, and describe what kind of star it will eventually become.

Name: _____

Course: _____

Section: _____

Remains of the Day

Introduction and Purpose

In the exercise "Old Age and Red Giants," we saw how a star becomes a red giant as it uses up the store of hydrogen in its core. In the exercise "Puff-Balls and Smoke Rings," we looked at the final stages of small to average stars with masses of less than about 10 solar masses. In today's exercise, we will examine the demise of the really big stars (those with masses greater than 10 times that of our Sun), and in doing so we will discover the remains of one of the most violent events in the universe—a supernova explosion.

The purpose of today's exercise is to examine the remains of some very large stars that have undergone supernova explosions. As in the exercise on planetary nebulae, we will be looking for common threads in the shapes, colors, and symmetries of these supernova remnants.

Procedure
Section 25.1

Start *RedShift College Edition*. Click on the **Go** button in the navigation bar, select **Science of Astronomy**, **Stars**, and **Birth and death of stars**, and click on page **35**. Watch and read through page **49** to discover what happens to very large stars when they run out of fuel.

Recall that as a star uses up its hydrogen, helium accumulates in the core to take the hydrogen's place. Over the millennia, this helium core gets more and more massive, and gravity crushes it smaller and smaller. Eventually, gravity will crush its core to the point where the helium will begin to fuse. At this point, the star has a new lease on life (at least temporarily). It usually blows away the outermost cool gas at its surface (perhaps forming a planetary nebula) and goes back to being a hotter, bluer star for a time.

If the star has an even larger mass, it will fuse the carbon core that results from the fusion of the helium. Again, the star will temporarily heat up and get bluer till all the carbon is gone. The largest stars can go through several such cycles of fuel burning, building up several internal layers of different elements (like the layers of an onion), but the outer, visible layer is always made of hydrogen.

25

Eventually, these huge stars will fuse silicon. When a star arrives at the point where it is fusing silicon, the fusion by-product is iron. Iron is the most stable nucleus known—and that spells trouble for this massive star. As the iron core forms, it collapses under its own weight, too, and the fusion of iron begins. Unfortunately for the star, iron fusion absorbs heat instead of releasing it, and the star's core, which has been creating light and heat since the star's very birth, suddenly goes out.

With no heat left in the core to push the star's material outward, gravity wins in an instant. The entire mass of the star falls inward in one gigantic crunch. The mass falling in from all sides compresses the mass in the very center till it cannot be compressed any more. At this point, the core becomes inflexible, and all the material that is still falling inward bounces right back out into space. It is like dropping two super-balls on top of one another. The combined energy of both balls is imparted to the upper one, and it rebounds. So the outer star matter is flung back out into space in a *supernova* explosion—a ragged splattering of gas and dust that will continue to expand and glow as a beautiful nebula for many millennia into the future.

What is left at the center is a tiny sphere of virtually solid matter. The atoms have been squeezed so tight that they have collapsed into one another. There are no more spaces between the nuclei, nor even any nuclei—just an ultra-tight packing of neutron-type material. All the protons have been flung away or transformed, and all the remaining electrons are spinning around on the surface of this tiny star. And the star is spinning fast now! Just as an ice skater spins faster and faster as she pulls her arms and legs toward the center of her spin, so the star spins much faster when its mass is pulled inward.

Can we see such rapidly spinning stars? It depends on the mass of the core that is left and on its orientation with respect to Earth. If the core has a mass larger than 3 solar masses, we will not see it. The gravity of the star will be so strong that it will prevent even light from escaping, and the star will have become an invisible *black hole*.

If the neutron star is smaller than 3 solar masses, however, its light will come out in a beam like a flashlight along its magnetic axis that will be rotating. It would be like laying a flashlight on a table and spinning the flashlight. If the direction of the beam happens to be aimed toward Earth at any point during the star's rotation, we will see it for a moment and then

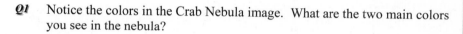

it will rotate away, only to reappear again for a moment at the next rotation. The result would be a star that blinked rapidly on and off in sync with the neutron star's fast rotation. In recent years we have been able to observe many of these stars, called *pulsars*.

Section 25.2

Click on the **Science of Astronomy** tab, and select **Photos** to open the Photo Gallery window. If the Contents frame is not open, click on the **Contents** button in the navigation bar to open it. In **The Galaxy** section of the Photo Gallery contents, select **Supernova remnants**. Scroll down to the image titled **[1067] Crab Nebula,** and click on the small icon to enlarge the image. This photograph is a real-color photo. Notice the colors in the image, the symmetry and the stars. While looking at this picture, answer the following questions.

Q1 Notice the colors in the Crab Nebula image. What are the two main colors you see in the nebula?

Q2 Look at the overall symmetry of the Crab Nebula. Does it appear to have exploded
    A.   equally in all directions?
    B.   randomly in all directions?
    C.   symmetrically along an axis?

Q3 Notice the twisty filaments of glowing gas in the Crab Nebula. Are they brighter than the rest of the nebula, or dimmer?

Q4 Given your answer to the last question, do you think these filaments are hotter or cooler than the rest of the material?

Q5 Now look at the stars in the image. Most of these stars are either in front of the nebula or behind it, but one of them is the remains of the star that blew up to form the nebula. Guess which star it is? (*Hint:* look for a fairly bright star near the geographic center of the nebula.)

25

In the space below, draw a rough sketch of the nebula showing the position of the star you have chosen.

Q6   In the sketch above, indicate the direction (or directions) along which you think the greatest force of the explosion traveled.

Section 25.3

Click the Thumbnail icon, and scroll to the image titled **[0594] Cassiopeia A: radio image**. Enlarge the image by clicking on the Magnify icon. Notice the symmetry of the nebula and the brightness of its various parts. This is a false-color image created with a radio telescope so that the stars (which do not emit many radio waves) do not show up. With the picture still on the screen, answer the following questions.

Q7   Compare the image of Cassiopeia A to the previous image of the Crab Nebula. What similarities do you find between the two?

Q8   Again, compare the image of Cassiopeia A to the previous image of the Crab Nebula. What differences do you find between the two?

Q9   In the image of Cassiopeia A, can you find a brighter spot of radio emissions where you think the neutron star ought to be?

Section 25.4

Click on the Thumbnail icon to go back to the list of images, and then scroll to the image titled **[0259] Part of the Vela supernova remnant**. Enlarge the image. Again, study the image's colors, symmetry and stars. With this image still on the screen, do the following:

Q10 List the features you see in the Vela supernova remnant that are similar to those in the previous two supernova remnants.

Q11 Do you see any sort of symmetry in this image? If so, briefly describe it.

Q12 Do you think you could locate the neutron star that caused this explosion?

Q13 In the two real-color photos (the Crab Nebula and the Vela supernova remnant), what color are the brightest filaments?

Q14 Since these filaments are made of gas that was near the outside of the star that exploded, what kind of gas would it probably be?

Section 25.5

Now bring the *RedShift 3* screen to the front, and open your personal settings file. Click on the Filters button, and select **Constellations**. Click in the **Boundaries** box to turn on the constellation boundaries, and then remove the check mark from the **Patterns** box to turn the patterns off. Now click on the **Deep Sky** tab. In the Deep Sky filters window, pull the upper right-hand **Magnitude** slider to **12** so the fainter nebulae in today's lab will show up, and then click **OK**. Change the sky settings as shown in the table below, and press the <Enter> key on your keyboard.

Local Date	Today's date
Local Time	9:00 p.m.
Zoom	10

Now that the sky is set correctly, let's find the Crab nebula. Select **Find** from the Information menu to open the Find window. In the **Find:** box, select **Deep Sky** and **Proper names**. Scroll down and double-click on **Crab nebula**, and then click **OK** to center the screen on the Crab nebula. It should be visible as a tiny, fuzzy patch in the center of the screen. Click on it to open its name tag, and then click on the name tag to open the Crab nebula's information box.

25

In the information box, click on the Report button. Finally, in the Report window, find the times of rising and setting at your viewing location for today's date, and record these in the data table. If the object is not visible on this date from your location, you will see the phrase **below horizon**. Now close the information box, find out what constellation the supernova remnant is in, and record these data in the data table below. Repeat the above steps to find and record the data for the Veil Nebula.

Q15 Which of these nebulae would be easily visible *from your location* at 9:00 p.m. tonight?

Data Table

Supernova Remnants							
Object	Constellation	Rise	Set	Object	Constellation	Rise	Set
Crab Nebula				Veil Nebula			

Section 25.6

Now insert the Extras CD in the computer's CD-ROM drive. When the *RedShift College Edition* startup screen appears, click **Start**. Then, when the Extras title screen opens, click on **Story of the Universe** and choose the section titled **Lives of the Stars**. You have already viewed the first part of this narrated animation in the exercise on red giants, so you may click just to the right of the center of the position bar if you wish to skip the first part. Pay special attention to the sections on planetary nebulae and supernovae, and then answer the following questions.

Q16 According to the animation you just watched, a planetary nebula will form from stars up to how many times the mass of our Sun?

Q17 When a planetary nebula forms, what will be at its center?

Q18 What will eventually happen to a white dwarf star?

Q19 In a supernova explosion, how long does the collapse of the core of the star take?

Q20 In a supernova explosion, what will be the *minimum* mass of the final, collapsed core of the star? (You will have to get this from your textbook.)

Web Intrigues

(If you do not have Internet access, you may skip this section.)

Log on to the Internet and open your web browser. Type in the following URL address: *http://chandra.harvard.edu/photo/0052/index.html*. This is the Crab nebula page of the Chandra orbiting X-ray observatory, launched in 1999. On this page you can see what this famous supernova remnant looks like in several different wavelengths. Click on the *Jpg* links below each photo to see it full size. After viewing the images, click on the *Caption* link to read about the Chandra X-ray image. In the space below, describe and compare the different views.

Now type in the following URL: *http://oposite.stsci.edu/pubinfo/pr/1996/22.html*. This is the press release page for the Hubble Space Telescope's time-lapse movie of the Crab Nebula. First, click on the **Press Release Text** link to discover what astronomers learned from the HST's observations of the Crab nebula pulsar. If you have the time, scroll down to the link titled **Crab Nebulae—The Movie**, download it, and watch it. Then answer the following questions.

Q21 How large in diameter do astronomers estimate the Crab nebula pulsar (rotating neutron star) to be?

Q22 How fast does this star rotate?

Discussion

In these images of supernova remnants we have seen a variety of different outcomes from the same kind of event. We saw very symmetrical, spherical shells of gas and dust in the Cassiopeia A remnant, somewhat symmetrical bilateral lobes of material in the Crab Nebula, and (at first glance) asymmetrical arcs of material in the Vela supernova remnant.

Both the Crab Nebula and the Vela supernova remnant have pulsars at their geographic centers, but so far no pulsar has been found in the Cassiopeia A remnant. The Crab Nebula pulsar was actually the first pulsar to be discovered and is the upper (slightly dimmer) of the two bright stars near the center of the nebula. As it blinks on and off 11 times per second, you would never notice it unless you photographed it with a high-speed camera.

25

Conclusion

In the space below, summarize the main visible features of a supernova remnant. Describe what variations (the different forms) might be seen, and describe what connection pulsars have with supernova remnants. Discuss the initial conditions needed to produce a supernova explosion rather than a planetary nebula. Finally, describe the progressive stages a star goes through from the main sequence to its final end for (1) an average star like our Sun and (2) a very massive star. (Continue your conclusion on a separate sheet of paper if necessary.)

Name: _____

Course: _____

Section: _____

Dust Balls and Gas Bags

Introduction and Purpose

Have you ever started cleaning your house or apartment and wondered how it could have gotten so dusty in such a short time? In our world, dust seems to be one of the natural consequences of daily living. Well, take comfort—it's that way in the universe, too! The universe is littered with dust balls and huge clouds of gas, and it's a good thing! It was a huge cloud of dust and gas that gave birth to our Sun and Earth.

In today's exercise, we wish to examine some of these concentrations of dust and gas and try to discover some of their characteristics. The word **nebula** was coined many years ago to categorize these glowing fuzzy regions in the sky. It simply means "without specific shape." Some nebulae do turn out to have specific shapes or symmetries like spheres or hourglasses, but most do not. For the sake of simplicity, we will subdivide the nebulae into four categories: (1) emission nebulae, (2) reflection nebulae, (3) absorption nebulae, and (4) the remains of old stars.

We have already looked at this last group of nebulae, the remains of old stars, in the exercises "Puff-Balls and Smoke Rings" and "Remains of the Day." The symmetric spherical shells of gas were called *planetary nebulae*, and they were the remains of small to middle-size stars that had finished fusing their hydrogen. The larger, more ragged splatters of gas and dust were the remains of supernova explosions of very large stars.

In today's exercise, we will examine the other three types of nebulae. While the fourth type came at the end of a star's life, the first three are usually associated with its beginning. Now that we know what a star looks like when it dies, we are going to learn what it looks like before it is born.

Procedure
Section 26.1

Start *RedShift College Edition*. Click on the **Science of Astronomy** selection tab, and select **Photos**. If the Contents frame is not open, click the **Contents** button in the navigation bar to open it. In the Photo Gallery contents, find the heading **The Galaxy**, and select **Interstellar matter** to put a list of interstellar matter photos in the Photo Gallery window.

26

In the Interstellar matter photo list, move the scroll bar all the way to the bottom. The next-to-last image is **[0158] Trifid Nebula in Sagittarius (Messier 20)**. Click on the small Magnify icon on the picture to enlarge it. In this image, you can see all three of the types of nebulae we will study in this exercise: emission, reflection and absorption. Spend a moment studying this picture, and then answer the following questions.

Q1   What are the two main colors you see in this nebula?

Q2   You have seen one of these colors before in the planetary nebulae and the supernova remnants. What gas caused it?

Q3   This gas (see the previous question) is glowing because it is being hit by high-speed, electrically charged particles. These particles had to have been accelerated by something pretty powerful, such as a large, hot star. Do you see any such stars in the center of this nebula?

Q4   The glowing red/pink gas is called an *emission nebula* because it actually emits light. The blue color is caused by light being reflected off small particles of dust. It is called a *reflection nebula*. Remember the color of this reflected light. What is the most probable color of the star causing this light?

Q5   Would a blue star be the sort of star to fit the description given in Q3?

Q6   The third type of nebula in this picture, the *absorption nebula*, is more subtle. There is one other color running through the pink and the blue in this nebula (or perhaps we should say "lack of color"). What is it?

Q7   Can you see background stars shining through the emission and reflection parts of the nebula?

Q8   Can you see background stars shining through the dark streaks trisecting the emission part of the nebula?

Q9   What can you deduce about the nature of those dark streaks from your answers to the previous two questions? Are they just gaps in the gas cloud where there is no material, or are they something else?

Q10 Look closely at the star-field around the nebula. Can you see any other dark areas in the star-field that seem to have fewer stars?

Section 26.2

Click the Thumbnail icon to return to the Interstellar matter section of the Photo Gallery. Scroll almost all the way to the top, and enlarge the image titled **[0238] The stars that excite the Trifid Nebula**. Answer the following questions.

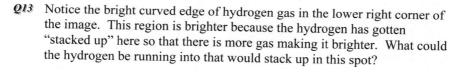

Q11 This image is a close-up of the center of the Trifid Nebula. Here you can see what lights up this beautiful nebula. How many stars are at its center?

Q12 Here, too you can get a better perspective on the dark streaks or lanes, as they are called, that transect the bright nebula. Let your brain go into 3-D mode as you look at the lane to the right of the central stars. Notice the extreme density of the dark material just to the right of the center. Think of it as a black cloud of sooty smoke floating above the twisting gaseous canyon that emerges toward the lower right. Can you perceive depth here?

Q13 Notice the bright curved edge of hydrogen gas in the lower right corner of the image. This region is brighter because the hydrogen has gotten "stacked up" here so that there is more gas making it brighter. What could the hydrogen be running into that would stack up in this spot?

Section 26.3

Click the Thumbnail icon to return to the Interstellar matter section of the Photo Gallery. Scroll down and enlarge the image titled **[0252] The dust lanes in Messier 16**. This is the Eagle Nebula, made famous in 1995 by the spectacular Hubble Space Telescope images of the "Elephant Trunks" of dust reaching up from the center of the nebula. Study the many dark "fingers" or "lanes" of dust that lace this vast cloud of glowing hydrogen. Then answer the following questions.

Q14 Are the pockets of dense dust in the Eagle Nebula more concentrated and well-defined than in the Trifid Nebula, or less so?

Q15 Would you say there is more dust in the Eagle Nebula than in the Trifid, or less?

26

Q16   Do you see more bright stars in the Eagle or in the Trifid Nebula?

Q17   Since stars are mostly made of hydrogen, what would you guess about the number of stars that might be formed from such a cloud as the Eagle Nebula?

Section 26.4

Click the Thumbnail icon to return to the contents of the Interstellar matter section of the Photo Gallery. Scroll down and enlarge the following images, one at a time:

[0159] The Lagoon Nebula, Messier 8
[0260] NGC 6559 and IC 1274-5, nebulosities in Sagittarius
[0264] The Rosette Nebula and NGC 2244

Now answer the following questions.

Q18   After looking at these different nebulae, what would you guess is the most common type of gas in our universe?

Q19   List as many features as you can that were present in all of these images.

Section 26.5

Click the Thumbnail icon to return to the contents of the Interstellar matter section of the Photo Gallery. Now scroll to the image titled **[0149] The Orion Nebula, M42 and M43**. Enlarge the image, and answer the following questions.

Q20   The Orion Nebula is one of the most incredible objects in the sky. The visible nebula is around 15 light years across and contains not only enormous quantities of gas but also a huge molecular cloud visible only in radio telescopes.

The visible emission nebula is powered almost entirely from a tiny cluster of stars at the nebula's very center. It is called the *Trapezium*. Look closely at the picture, and count the stars in this tiny group. How many are there?

Q21   In this picture, what other color do you see besides the red of the hydrogen gas?

Q22 Return to the Interstellar matter page and read the information describing this image. What gas is causing this color? (See the previous question.)

Section 26.6

Click the Thumbnail icon to return to the Interstellar matter photo list and scroll to the very bottom of the page. Enlarge the image titled **[3071] A region of recent star formation in the Orion Nebula (HST)**, and answer the following questions.

Q23 This image is an extreme close-up of the central area of the Orion Nebula. Study the upper central area of the picture. If you look carefully, you can see a couple of tiny orangish-yellow "globules" of material suspended in front of the rest of the nebula. Look very closely at the largest elliptical one in the center of the image. (Use a magnifying glass if needed.) What do you see in the center of this globule?

Q24 How many other, similar globules can you count in this image?

Q25 If a star is born out of a cloud of gas and dust in which gravity has concentrated the material in the cloud's center, would it look like these globules?

Q26 If you could look at this region in other wavelengths besides visible light, in which wavelength would you want to look to see if there is a baby star forming in the center of these globules? (Would a baby star be hotter or cooler than a regular star that gives off visible light?)

Section 26.7

In the Contents frame, select **Unusual stars** under the heading **The Galaxy**. Scroll down the Unusual stars photo list to the image titled **[3063] Young stars in the Orion Nebula**, and enlarge it. Answer the following questions.

Q27 This image is an extreme close-up of the previous one. The elliptical globule in the center is the one you just examined. You can clearly see the dark concentration of matter at the center of this cocoon of material. Besides the other bright globules, do you see any dark globules that could be accretions of dense matter forming around a proto-star?

Q28 Get a magnifying glass and magnify the small, dark, circular spot at the left of the picture. What color do you see at its center?

26

Q29 This is one of the most exciting baby pictures of a star we have ever taken. This spherical ball of dense, dark matter is cloaking a tiny, hot center—a star about to be born. What will the dark, dense material around the new star turn into if gravity continues to condense it into smaller objects?

Q30 There is one other interesting thing about this picture. What similarity can you find in the shapes of the two larger bright globules of gas and dust?

Q31 What appears to be happening to the outer regions of the gas around these globules?

Q32 If the gas in a nebula is glowing, it is because it is being lit up by charged particles flying away from some nearby hot, bright star. Given your answers to the previous two questions, in which direction would you guess the nearby star is? (Look back at the previous picture [it is reversed from this one] and you can identify the culprit.)

Section 26.8

In the Contents frame, select **Interstellar matter** again. Now scroll through the Interstellar matter photos to the image titled **[0258] The Horsehead Nebula and NGC 2024 in Orion**. Enlarge the image, and answer the next question.

Q33 Study this image carefully, and then summarize the following information in the Conclusion section of this exercise. List all the different kinds of nebulae you see here. List all the different colors you can see in these nebulae, and tell what causes each color. List all the different features related to these nebulae that are visible in this image (such as bright shock fronts, streamers of gas, dark globules, etc.). Based on what you have learned, would you expect new stars to be born here in the future?

Section 26.9

Click on the selection tab (currently set to **Photos**), and select **Science of Astronomy**. In the Contents frame, select the segment titled **Birth and death of stars** from the **Stars** chapter. Read and watch through page **17** for a good summary of today's exercise.

Section 26.10

Now bring the *RedShift 3* screen to the front, and open your personal settings file. Click on the Filters button and select **Constellations**. Click in the **Boundaries** box to turn on the constellation boundaries and then remove the check mark from the **Patterns** box to turn the patterns off. Now click on the **Deep Sky** tab. In the Deep Sky filters window, pull the upper right-hand **Magnitude** slider to **12** so the fainter nebulae in today's lab will show up, and then click **OK**. Change the sky settings as shown in the table below, and press the <Enter> key on your keyboard.

Local Date	Today's date
Local Time	9:00 p.m.
Zoom	10

Now that the sky is set correctly, let's find the Eagle Nebula. From the Information menu select **Find** to open the Find window. Click on the small arrow to the right of the **Find:** box, select **Deep Sky**, and click on **Proper names**. In the list box, double-click on **Eagle Nebula**, and click **OK** to center the screen on the Eagle Nebula.

In the information box, click on the Report button. Finally, in the Report window, find the times of rising and setting at your viewing location for today's date, and record these in the data table. If the object is not visible on this date from your location, you will see the phrase **below horizon**. Now close the information box, and set the Zoom to **1**. Find out what constellation the supernova remnant is in, and record this data in the data table below. Repeat these steps to find and record the data for each of the remaining nebulae in the data table below.

Q34 Which of these nebulae would be easily visible *from your location* at 9:00 p.m. tonight?

Data Table

Emission Nebulae							
Object	Constellation	Rise	Set	Object	Constellation	Rise	Set
Eagle Nebula				North American Nebula			
Horsehead Nebula				Omega Nebula			
Lagoon Nebula				Orion Nebula			

Web Intrigues

(If you do not have Internet access, you may skip this section.)

Log on to the Internet and open your web browser. Type in the following URL: *http://oposite.stsci.edu/pubinfo/1999.html*. This page holds the index of Hubble Space Telescope images released in 1999. Scroll down and select **PR**

26

99-20 Hubble Snapshot Captures Life Cycle of Stars. When the new page has opened, read the text below the image. When you think you understand the significance of each part of the picture, click on the image itself to open a page with **JPG** icons, and then click on the first one to enlarge the image.

Q35 List the four stages of a star's life that are on display in this amazing image.

Now type in the following URL: *http://www.ast.cam.ac.uk/AAO/images.html*

This is the Anglo Australian Observatory's photo gallery page. On the left of the page, click on the **Icons** link beside the title **Emission Nebulae**. This will take you to a page featuring a whole gallery of incredible photos of star-birth regions all over the sky. Look at as many of these beautiful images as you wish. Then scroll down to the **Scorpius** section, and click on the image **NGC 6334**. Look at this beautiful nebula and read the accompanying text. Then answer the following questions.

Q36 What is so unusual about this particular nebula?

Q37 What is the cause of this different coloration?

Q38 How far away is this glowing cloud of gas and dust?

Discussion

When we look across our galaxy (and into other galaxies), we see lots of the raw material from which stars are made in huge glowing clouds called nebulae. Most of this visible material is glowing because of hot, young stars that recently formed within these clouds. These nebulae also contain objects that fit the description of what we would expect from stars in various stages of birth. We see these objects in infrared images. We see globules that appear to be the cocoons of them, and we even see disc-shaped proto-solar-systems of dark material surrounding them.

Conclusion

On a separate sheet of paper, write a short conclusion, summarizing the items listed in question Q33.

IV
Groups and Galaxies

Contents

Name: _____

Course: _____

Section: _____

Double Stars

Introduction and Purpose

There are well over 100 stars brighter than 3rd magnitude as seen from Earth. Of these bright stars, many are in reality multiple-star systems rather than isolated stars. As astronomers have surveyed the stars in the Milky Way galaxy, they have concluded that single stars like our Sun are actually in the minority. More than half the stars we have observed either have one or more companions or are part of a large group of stars that are gravitationally connected with one another.

There are many close pairs of stars that could be called *double stars*. Some of them are really quite close together and are held together by their mutual gravity. These we call *binary stars*. Others are just accidental alignments of stars that look close together but are really very far apart and not related to each other at all. These we call *optical double stars*. It is the binary stars we are interested in today. In today's exercise, we will take a look at the brighter binary stars we can see from Earth. Most of those we will investigate today are easily resolved as double stars in small telescopes. Some are even resolvable in binoculars. All of them, however, appear to be single stars to the naked eye.

Procedure

Section 27.1

Start *RedShift College Edition*, bring the *RedShift 3* screen to the front, and open your personal settings file. Click on the Filters button and select **Constellations**. In the Constellations filters window, click in the **Boundaries** box, and click **OK** to turn on the constellation boundaries. Change the sky settings as shown in the table below, and press the <Enter> key on your keyboard.

Local Date	**Today's date**
Local Time	**9:00 p.m.**
Zoom	**10**

Now, look at the data table on page 269, and you will see the names of some of the brighter binary stars in the sky. To begin today's exercise, use *RedShift's* **Find** feature to locate each of the stars in the data table and fill in the empty columns. First, select **Find** from the Information menu. In the Find window, click on the small arrow to the right of the **Find:** box, select **Stars**, and click on **Proper names**. Now scroll down through the Proper names list box, double-click on the star you wish to investigate, and click **OK** to center the screen on the star.

Click on the star in the center of the screen to open its name tag and verify that it is the correct star. Look at the screen to discover the constellation the star is in, and write that in the first column of the data table on page 269. (You may need to temporarily reduce the Zoom to **1**.) Now click on the name tag to open the star's information box, and then click on the **More** button to expand the box. Finally, click on the **Structure** tab to see the relationship between the two stars in the binary system.

Find and record the number of components in the system and the Angular separation of the two stars. (Use an average if the 1991 and 1999 values are different.) Then find the Position angle in 1991 and the Position angle in 1999, and record these data in the appropriate columns of the data table.

To fill in the Estimated Period column of the data table, you must do a few calculations. First, subtract the position angle in 1999 from the position angle in 1991 (ignore any minus sign), and record this number in the Change in Angle column. This is the number of degrees star B moved around star A in 8 years, so divide 8 years by the Change in Angle to find the number of years it takes the star to move one degree.

$$\text{Years/degree} = \frac{8 \text{ years}}{\text{Change in Angle}}$$

The period of the system is how many years it takes for star B to go 360 degrees all the way around star A. To get this number from your data, multiply the number of years/degree (your previous answer) by 360 degrees.

$$\text{Period} = \frac{\text{Years}}{\text{Revolution}} = (360 \text{ degrees/rev}) * (\text{Years/degree}) = 360 * (\quad)$$

Record this answer in the Estimated Period column.

Repeat this process until you reach the last two stars in the data table. To find these stars, click the small arrow to the right of the **Find:** box, select **Stars**, and click on **Bayer-Flamsteed**. These two stars are listed as **Eps_1 Lyr** and **Eps_2 Lyr**. They are about halfway down the list. When you finish, answer the questions below.

Data Table

Bright Binary Stars							
Star Name	Constellation	Number of Components	Angular Separation	Position Angle (1991)	Position Angle (1999)	Change in Angle	Estimated Period
Acamar	**Eridanus**	2	8"	90	91	1	2880 y
Achird							
Algeiba							
Albireo							
Dschubba							
Dubhe							
Graffias							
Rigel Kentaurus (+Proxima)		+1					
Rotanev							
Sabik							
Tegmen							
Epsilon 1 Lyrae							
Epsilon 2 Lyrae							

Q1 The Epsilon Lyrae system includes both Epsilon 1 Lyrae and Epsilon 2 Lyrae, and the Rigel Kentaurus system includes Proxima Centauri. Thus, we have 12 star systems in this data table. Add all the numbers (including the +1 for Proxima Centauri) in the Number of Components column to find out how many individual stars are represented in these 12 star systems.

Q2 Calculate the average number of individual stars per star system in this sampling by dividing your answer to Q1 by 12.

Average number of stars/system = (**Q1**)/12

=

Q3 Though it is not listed as such in the *RedShift* database, Sirius is also a double star; it has a white dwarf companion. In a star system like Sirius,

27

where one star is a main sequence A star, and the other is a white dwarf, which star has already used up most of its fuel?

Q4 Assuming that both stars were born at the same time, which star used its fuel the fastest?

Q5 Given one star that is larger than another, which star will be the hottest?

Q6 Which star will use its fuel the fastest?

Q7 Given your answers to questions Q4–Q6, which of the two stars making up the Sirius binary star system was originally the largest and hottest, the main sequence A star or the white dwarf?

Q8 Examine the periods of these binary stars. The longer it takes two stars to orbit one another, the farther apart they must be. On the basis of this information, which pair must have the greatest actual separation from one another in space?

Q9 Now look at the Angular Separation for each pair. Does the pair with the largest actual separation distance (your answer to Q8) also have the largest apparent angular separation (from the data table)?

If stars that really are far apart (have large separation distances) seem to be close together, they must be far away.

Q10 On the basis of your answer to Q9, which binary star system is probably the farthest away? (Look for one with a long period and a small angular separation.)

Section 27.2

Now bring the *RedShift College Edition* window back to the front. Click on the **Go** button in the navigation panel, select **Science of Astronomy**, **The Astronomer's Tool Chest**, and **Messages in the shadows**, and click on page **14**. Read through page **18** to learn about two other types of binary star systems. Then answer the following questions.

Q11 What do we sometimes find in the spectra of a star that tells us it is a binary pair?

Q12 What do we call a binary star system where one star periodically passes in front of the other?

Q13 How do we detect binary systems like the ones described in Q12?

Web Intrigues

(If you do not have Internet access, you may skip this section.)

Log on to the Internet and open your web browser. Type in the following URL: *http://csep10.phys.utk.edu/astr162/lect/binaries/visual.html* This will take you to a page created by the University of Tennessee at Knoxville featuring a wonderful discussion of binary star orbits along with 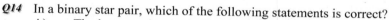 some really cool interactive applets that allow you to vary the different parameters of a binary star system and watch the results. Read through the page and try out the simulation applets. Then answer the following questions.

Q14 In a binary star pair, which of the following statements is correct?
A)   The less massive star orbits the more massive one.
B)   The more massive star orbits the less massive one.
C)   Both stars orbit their common center of mass.
D)   The orbit of each star is erratic and cannot be predicted.

Type in the following URL:
http://instruct1.cit.cornell.edu/courses/astro101/java/eclipse/eclipse.htm.
This is a page developed by Cornell University. Here you will find a very neat eclipsing binary simulator. You can vary the type of each star, the angle of inclination of their orbits, and the distance between the stars, and can see the effects on the light curve. (*Note:* After making any changes, you must click the **Enter values** button to see the effect of your changes.) After playing with this simulation for a while, use it to answer the following question.

Q15 If both stars in an eclipsing binary star are the same star type, which of the following statements is correct?
(A)   The dip in the light curve happens less often.
(B)   The depth of each dip is the same.
(C)   The depth of each dip becomes deeper.
(D)   The depth of each dip becomes very different.

Section 27.3

Now insert the Extras CD in the computer's CD-ROM drive. When the *RedShift College Edition* startup screen appears, click **Start**. Then, when the Extras title screen opens, click on **Story of the Universe** and choose the section titled **Double Stars**. Watch this narrated animation for a good review of today's topic, and then answer the rest of the questions.

Q16 What is the shape of the orbit of each star in a binary pair?

Q17 Explain why Algol gets brighter and dimmer.

Q18 What do we estimate as the original masses of the two stars in the Algol pair?

Q19 When double stars are really close together, what shape do they sometimes take?

Q20 How did the stars of Algol change over time?

Q21 Explain how a binary pair causes a nova.

Discussion

Binary stars are an important link in our chain of discovering how stars form. Since many stars form this way, we have come to realize that the original distribution of matter in our galaxy must have had many clumps that had a lot more matter than would be needed to make just one star. As a result, multiple-star systems formed simultaneously.

As you might expect, the distribution of matter in each large clump was not always perfectly uniform. Sometimes one star got more of the matter than the other and therefore was larger and hotter. These "piggy" stars use up their fuel rather extravagantly and die at a much earlier age than their more modest brothers. So often we encounter pairs that seem to be terribly mismatched— one star blazing away brightly while the other, now a decrepit, old white dwarf, is hardly noticeable.

Such unbalanced binary systems also lead to another remarkable astronomical phenomenon—the *nova*. As time passes, even the bright A star in Sirius will run out of fuel. As it does so, it will swell to become a red giant (see the exercise "Old Age and Red Giants"). At this point, its outer atmosphere (containing much of its unused hydrogen) will be puffed off its surface as a planetary nebula. Much of this hydrogen will be pulled onto the surface of the white dwarf companion. The white dwarf, though small, has a significant

27

gravitational pull. If enough hydrogen accumulates on the surface of the white dwarf, the temperature and pressure will be sufficient to renew hydrogen fusion, and the white dwarf will suddenly blaze forth brilliantly with renewed vigor.

Unfortunately, this will be short-lived as the hydrogen is quickly used up, and the white dwarf must fade away again to wait for more fuel from its now-aging brother. The visible result of this process is a usually faint star suddenly becoming bright for a few days and then fading back into obscurity. This is what we call a *nova*.

In some binary star systems, the red giant phase of the second star may last for a long time, leading to repeated episodes of nova activity. In some very extreme cases, the white dwarf may have almost enough mass to collapse to become a neutron star. Then an influx of matter from its red giant brother may push it over the edge, resulting in a tremendous supernova explosion.

Conclusion

In a few paragraphs, summarize what you have learned about binary stars. Include comments about how common binary stars are, about the variety of star types involved, and about how pairing different sizes of stars can lead to various astronomical phenomena. Also comment on the periods of binary stars and on their angular separations as seen from Earth.

Notes

27

Name: _____

Course: _____

Section: _____

Open Clusters

Introduction and Purpose

In the previous exercise, we saw that stars often form in pairs or even groups of four to ten. In today's exercise, we will examine some even larger groupings of stars called ***open clusters***. Some of the most notable groups of stars in the sky belong to open clusters. Two in particular are easily recognized, the Pleiades (better known as the "Seven Sisters") and the Hyades (the "V" of stars that form the face of Taurus, the bull).

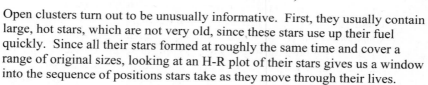

Open clusters turn out to be unusually informative. First, they usually contain large, hot stars, which are not very old, since these stars use up their fuel quickly. Since all their stars formed at roughly the same time and cover a range of original sizes, looking at an H-R plot of their stars gives us a window into the sequence of positions stars take as they move through their lives.

Second they are often associated with large, gaseous nebulae where stars are still forming. This fact allows us to compare the formation conditions in such places with the types of stars that are produced there. Finally, they turn out to be an excellent indicator of the spiral structure of most spiral galaxies like our own Milky Way. In today's exercise, we will look at several open clusters and observe some of these distinguishing features.

Procedure
Section 28.1

Start *RedShift College Edition*. If the Contents frame is not open, click on the **Contents** button in the navigation bar to open it. Now click on the **Science of Astronomy** tab, and select the **Photos** option. In the Photo Gallery contents, scroll down to the heading **The Galaxy**, and select **Open clusters** to put a list of open cluster photos in the Photo Gallery window.

In the open clusters photo list, scroll down to the image titled **[0272] The Reflection Nebula in the Pleiades Cluster**. Click on the small icon on the picture to enlarge it. This is one of the most famous and easily observed open clusters in the sky. Answer the following questions.

Q1   What is the predominant color seen in the stars of the Pleiades?

28

Q2   Does this color indicate a cool star, a medium-temperature star, or a hot star?

Q3   Count all the stars in this picture with spikes sticking out of them. (The spikes are produced when a very bright star is imaged by a Newtonian telescope. They are not actually part of the star.) How many are bright enough to produce the diffraction spikes?

Q4   What do you think is causing the wispy features around the brightest stars?

Q5   Only the six brightest stars in this cluster are visible to the naked eye. Yet the Greeks called the cluster the *Seven Sisters*. (It is known as the *Seven something-or-others* in a number of other mythologies as well.) Apparently thousands of years ago, seven stars were bright enough to be seen with the naked eye. Now that you have looked at a high-magnification image of the cluster, what reason can you give to explain the fact that today we can only see six bright stars?

Section 28.2

Click on the Thumbnail icon to go back to the contents of the Open clusters Photo Gallery. Scroll to the image titled **[0105] The Jewel Box cluster in Crux, NGC 4755**. Click on the small icon on the picture to enlarge it. Answer the following questions.

Q6   In the Jewel Box Cluster, what is the predominant color of the brightest stars?

Q7   In the Jewel Box Cluster, a couple of the very brightest stars have a different color. What is it?

Q8   Are these stars mentioned in Q7 hotter or cooler than those described in Q6?

Q9   If they are a different temperature, but the same brightness, what kind of stars must these oddballs (Q7 stars) be?

Q10 Are these stars (mentioned in Q7-Q9) closer to the end of their lives, in the middle of their lives, or near the beginning of their lives?

Q11 Assuming that all the stars in this cluster formed from the same cloud of dust and gas and that they are all actually about the same age, how can you explain your answer to Q10?

Q12 Do a quick count of the bright stars in the Jewel Box Cluster. Are there a lot more, a lot fewer, or about the same number of bright stars here as in the Pleiades star cluster?

Q13 Now look closely at the dim background stars in the Jewel Box Cluster image. What colors seem to predominate in these stars?

Q14 Comparing the stars making up the Jewel Box Cluster with those in the background, would you say the open cluster stars are average stars or extraordinary ones?

Section 28.3

Click on the Thumbnail icon to go back to the contents of the Open clusters Photo Gallery. In the contents frame, select **Interstellar matter** from **The Galaxy** section to change the Photo Gallery list. In the Interstellar matter photo list, scroll down about 5/6 of the way to the bottom to the image titled **[0246] The Rosette Nebula and NGC 2244**. Click on the small icon on the picture to enlarge it. Answer the following questions.

Q15 What is the predominant color in the Rosette Nebula?

Q16 What kind of gas makes this color?

Q17 What is the predominant color of the brightest stars in the center of the nebula?

Q18 Are there any bright stars with the second color (see Q7) noted in the Jewel Box Cluster?

Q19 Return to the Interstellar matter image list, and read the introduction to this picture. Tell why the center of the nebula is fairly free of gas.

28

Q20 As time goes on, what will probably happen to the rest of this gas?

Q21 What do you think this region of the sky will eventually look like?

Q22 How many of the stars in these open clusters appear to have progressed off the main sequence?

Q23 On the basis of your answer to the previous question, what can you say about the age of stars in open clusters?

Q24 What can you say about the amounts of dust and gas present in the open clusters you looked at today?

Section 28.4

Now bring the *RedShift 3* screen to the front, and open your personal settings file. Click on the Filters button and select **Constellations**. In the Constellations filters window, click in the **Boundaries** box and click **OK** to turn on the constellation boundaries. Change the sky settings as shown in the table below, and press the <Enter> key on your keyboard.

Local Date	Today's date
Local Time	9:00 p.m.
Zoom	3.5

Now that the sky is set correctly, let's find the Pleiades. From the Information menu, select **Find** to open the Find window. Click on the small arrow to the right of the **Find:** box, select **Deep Sky**, and click on **Proper names**. In the list box, double-click on **Pleiades**, and click **OK** to center the screen on this beautiful open star cluster. It should be visible as a small group of bright stars in the center of the screen. Click on it to open the name tag, and then click on the name tag to open the Pleiades's information box.

In the information box, click on the **Report** button. Finally, in the Report window, find the times of rising and setting for today's date and record these in the data table. If the object is not visible on this date from your location, you will see the phrase **below horizon**. Now, select **1** from the drop-down Zoom menu, and close the information box. Find out what constellation this cluster is in and record that in the data table. Now select **3.5** from the drop-down Zoom menu. Repeat the above steps to find and record the data for each of the remaining open star clusters in the data table. (For the last one, M6, you will need to change the **Find:** box to **Deep Sky, Messier numbers**.)

Q25 Which of these open star clusters would be easily visible *from your location* at 9:00 p.m. tonight?

Data Table

Open Clusters							
Object	Constellation	Rise	Set	Object	Constellation	Rise	Set
Double Cluster in Perseus				Pleiades			
Beehive				Wild Duck Cluster			
Jewel Box Cluster				M6			

Web Intrigues

Bring the *RedShift College Edition* screen back to the front. Click the selection tab (currently set to **Photos**), and select **Web** to open your browser window. Click on the image in the browser window to open the *RedShift* web site. Once the *RedShift* web site has loaded, click on the folder titled **Our Galaxy**. On the Our Galaxy page, click on the **Star Clusters** folder, and on the Star Clusters page, select **Open star clusters**. On the Open star clusters page, click on the link **Messier Object 6**. This will take you to a page featuring a short description of one of the open star clusters in the direction of the center of the Milky Way, M6. Read the text, and then click on the link **NOAO color photo of M6**. When the new page loads, click on the link **Hi-res version of this image** to see a stunning color photograph of this beautiful open star cluster. Study the colors of the stars in this photo, and then answer the following questions.

Q26 Are the types and colors of the stars in M6 similar to those seen in other open clusters?

Q27 What color (or colors) are the brightest stars in this cluster?

Q28 Do blue stars usually burn their fuel slowly or quickly?

28

Q29   Since there are still so many of these stars (Q27, Q28) here, what can you say about the age of this cluster?

Q30   There are many red stars in this photo (a rarity in most stellar photos). Are these stars bright or dim?

Q31   These are not red giant stars, or else they would be very bright, so they must be main-sequence stars. What can you deduce about the mass and size of these stars? Are they larger than, smaller than, or about the same size as our Sun?

Discussion

As you have seen in these photographs, open clusters contain tens to hundreds of bright stars. These clusters must have formed from very large clouds of dust and gas such as is seen in the Rosette Nebula. Because most of the stars in these open clusters are blue and therefore very hot, we deduce that they must also be very massive. This would imply that the cloud of matter from which the cluster formed was quite dense when the star formation began. Otherwise, there would not have been enough matter in the vicinity to have made such large stars.

Most of the stars in these clusters would probably have begun forming around the same time. So when we see a few individual stars that have already moved into the red giant phase of their life, we know that these stars must once have been even larger and brighter than those that are still on the main sequence. By plotting the stars in open clusters on the H-R diagram, we find a neat progression of stars scattered all along a small portion of the main sequence. We usually find a few that have already moved off the main sequence and sometimes one or two that have progressed all the way over to the red-giant region of the graph.

Conclusion

On a separate sheet of paper, describe the visible characteristics of an open cluster. Characterize the age and types of the stars in such clusters as well as the amount of dust and gas present in them. Discuss other deep-sky objects often associated with open clusters. Finally, mention some of the information we can glean by studying such clusters.

Optional Observing Project

Do the observation exercise "Double Stars and Star Clusters" in Appendix C.

Globular Clusters

Introduction and Purpose

In the last exercise, we examined the nature of large groupings of stars called *open clusters*. Today, we will take a look at even larger groups of stars called *globular clusters*. While the typical open cluster has dozens to a few hundred stars in it, the typical globular cluster has thousands to a few million stars.

As you look at the globular clusters in today's exercise, mentally compare what you see with the open clusters you examined in the last exercise. You will find that the difference in the number of stars is only the beginning of the differences between these two very different kinds of star clusters.

Procedure
Section 29.1

Start *RedShift College Edition*. If the Contents frame is not open, click the **Contents** button in navigation bar to open it. Now click on the **Science of Astronomy** selection tab, and select **Photos**. In the Photo Gallery contents, scroll down to the heading **The Galaxy**, and select **Globular clusters**.

Scroll down to the image titled **[1106] Omega Centauri in Centaurus**. Click on the small Magnify icon on the picture to enlarge it. This is one of the most magnificent and easily observed globular clusters in the sky. Answer the following questions.

Q1   With transparent tape, mark off a 1-cm-by-1-cm square on your monitor screen about one-fourth of the way from the center of the cluster to the edge. Using a magnifying glass or reading glasses, count the stars in this square. (Include faint blobs as well as bright ones.) Record the number here.

Q2   With a ruler, measure the approximate radius of the Omega Centauri cluster (as it appears on your monitor) from its center out to where the stars get sparse. What is its radius in cm?

$$r = (\quad) \text{ cm}$$

29

Q3 Using the following formula and the radius you calculated in Q2, calculate the area of the cluster (as it appears on your monitor).

$$A = \pi(r)^2$$
$$= \pi(\qquad)^2$$
$$= (\qquad) \text{ cm}^2$$

Q4 Now you can approximate how many stars are in the Omega Centauri globular cluster. Multiply your answer to Q1 by your answer to Q3 to get the approximate number of stars here.

Number of stars = (Number of stars/cm^2) * (Number of cm^2)
$$= (Q1) * (Q3)$$
$$= (\qquad) * (\qquad)$$
$$= (\qquad) \text{ stars}$$

Q5 What shape is Omega Centauri?

Q6 In which part of the cluster do the stars seem the closest together?

Section 29.2

Click on the Thumbnail icon to go back to the contents of the Globular clusters photo list. Scroll down to the image titled **[0121] The globular cluster 47 Tucanae (NGC 104)**. Click on the small Magnify icon to enlarge this photo, and answer the following questions.

Q7 The picture of the globular cluster 47 Tucanae is in color. What color are most of the stars?

Q8 Contrast the colors of the stars in this picture with the colors of the stars in the open clusters.

Q9 We can tell by their spectra that most of the stars in globular clusters are still on the main sequence and have not yet moved to the red giant stage. Remember that on the main sequence, the less massive a star is, the cooler and redder it is.

Since these main-sequence stars are yellow-orange or red, what can you deduce about their masses—are they more massive or less massive than most of the stars in open clusters?

Q10 If large, hot stars use their fuel much faster than small, cool stars, which clusters are the oldest, open clusters or globular clusters?

Section 29.3

Click on the Thumbnail icon to go back to the contents of the Globular clusters photo list. Scroll to the image titled **[0128] The inner part of the M5 globular cluster.** Click on the small icon on the picture to enlarge it, and answer the following questions.

Q11 In the picture of the inner part of the globular cluster M5, what is the predominant star color?

Q12 If you look closely you will see that a few of the brightest stars in the center are a different color. What color are they?

Q13 When you observe stars in a telescope, only the brightest stars show color because of the structure of the color sensors in our eyes. The dimmer stars will appear to be white or shades of gray. Looking at the color of the brightest stars in the center of M5, what color do you think the cluster would appear to have as seen by your eye through a telescope?

Q14 What do you think the night sky would be like if you lived on a planet circling a star near the center of a globular cluster?

Q15 When a pizza chef whirls a ball of pizza dough and tosses it into the air, what shape does it change into as it is spinning?

Great pizza chefs spin the dough to make it flatten out into a lovely, smooth pizza crust. The spinning action slings the less well-attached parts of the dough to the edge and pulls the heavier, lumpy dough from the center outward and toward the plane of the pizza crust at the same time. This feeds more and more dough into the flat outer regions while reducing the amount in the center.

Q16 Considering your answer to the previous question and the shape of the globular clusters, would you expect globular clusters to rotate very fast?

Section 29.4

Click on the **Go** button in the navigation bar, select **Science of Astronomy, Galaxies**, and **The Milky Way**, and click on page **7**. Read through page **11**, and answer the following questions.

Q17 How are Population II stars different from stars like our Sun?

Q18 Which kind of star predominates in globular clusters, Population I or II?

Q19 Now click on the **Go** button in the navigation bar, select **Science of Astronomy**, **Stars**, and **Birth and death of stars**, and click on page **35**. Read through page **37**. Where did the heavier elements found in Population I stars (like our Sun) come from?

29

As a star goes through its life, it is continually creating heavy elements in its interior. When that star dies, it seeds the space around it with many of those heavy elements either through the formation of planetary nebulae or in supernova explosions. When subsequent stars form in that area, they will be made from the material shed by the previous stars.

Q20 Population II stars, however, have almost no heavy elements. If heavy elements are formed as suggested in the previous question, how many stars could have existed in that area before the Population II stars formed?

Q21 Consider your answer to the previous question. If Population II stars were among the first stars to form in a galaxy, would they be young or old?

Q22 In question Q10, you deduced the age of the stars in globular clusters based on their masses and the average age for stars of that mass. In question Q21, you deduced the age of the stars in globular clusters based on their composition. Do your two different deductions yield the same answer?

Whenever you get the same answer through two completely different methods, you have a lot more confidence that it is right.

Section 29.5

Now bring the *RedShift 3* screen to the front, and open your personal settings file. Click on the Filters button and select **Constellations**. In the Constellations filters window, click in the **Boundaries** box and click **OK** to turn on the constellation boundaries. Change the sky settings as shown in the table below, and press the <Enter> key on your keyboard.

Local Date	**Today's date**
Local Time	**9:00 p.m.**
Zoom	**3.5**

Now that the sky is set correctly, let's find some globular clusters. From the Information menu, select **Find** to open the Find window. Click on the small arrow to the right of the **Find:** box, select **Deep Sky**, and click on **Proper names**. In the list box, double-click on **Omega Centauri**, and click **OK** to center the screen on this incredible globular star cluster. It should be visible as a round fuzz-ball in the center of the screen. Click on it to open its name tag and then click on the name tag to open Omega Centauri's information box.

In the information box, click on the **Report** button. Finally, in the Report window, find the times of rising and setting at your viewing location for today's date, and record these in the data table. If the object is not visible on this date from your location, you will see the phrase **below horizon**. Now close the information box, find out what constellation this globular cluster is in, and record these data in the data table below. (You may need to temporarily decrease the magnification to see the constellation name.)

For the remaining globular clusters, you will need to change the **Find:** box to **Deep Sky**, **Messier numbers**. Repeat the above steps to find and record the data for each of the remaining star clusters in the data table.

Q23 Which of these globular star clusters would be easily visible *from your location* at 9:00 p.m. tonight?

Data Table

Globular Clusters							
Object	Constellation	Rise	Set	Object	Constellation	Rise	Set
Omega Centauri				M5			
M3				M13			
M4				M22			

Discussion

Globular clusters are some of the most impressive deep-sky objects that can be seen through amateur telescopes. When the atmosphere is turbulent, the stars seem to move in and out of the cluster's center like bees on a hive. Living on a planet orbiting a star in an open cluster would not be terribly different from living on Earth. But living on a planet orbiting a star in or near a globular cluster would be a very different experience. Imagine watching a

29

ball of stars covering a tenth of the sky slowly rising over the horizon like an enormous full moon!

Globular clusters are as different from open clusters as night and day. In addition to the differences we have already noted by just looking at them, we find that they are made of different kinds of stars as well. When we analyze the spectra of the stars in open clusters and in globular clusters, we find that the stars in open clusters have significant amounts of heavier elements like iron, magnesium, calcium, and silicon in their atmospheres. Stars in globular clusters have almost none of these elements in their spectra.

We call the mostly hydrogen and helium stars found in globular clusters *Population II* stars, and the heavy-metal stars found in open clusters *Population I* stars. Our Sun is a Population I star. In the exercises about our Milky Way galaxy we will explain the reasons for most of these differences as well as discover that these two kinds of clusters help identify the different parts of a typical spiral galaxy.

Conclusion

Write a paragraph or two summarizing the characteristics of globular clusters. Compare and contrast these characteristics with those of open clusters. Include such things as the colors and shapes of the clusters, the types of stars making them up, and their ages. (Continue your conclusion on a separate sheet of paper if necessary.)

Optional Observing Project

Do the observation exercise "Double Stars and Star Clusters" in Appendix C.

The Milky Way:
The Spiral Arms

Introduction and Purpose

Anyone who has been lucky enough to be in the desert on a clear, starry summer night knows what an amazing sight the Milky Way presents. Because the galaxy is not bright, it is not seen in most urban and many rural areas. However, when the sky is not lit up by cities, towns, or the Moon, the Milky Way dominates the sky. It stretches from one end of the summer sky to the other, weaving in and out of the bright stars, making even familiar constellations difficult to pick out.

Today we have come to realize that the Milky Way is the central region of our own galaxy and that we are seeing it from somewhere inside. But for virtually all of history, this has not been the case. Only through connecting several apparent coincidences were astronomers able to piece together the puzzle of the shape of our galaxy and of our position in it.

In today's exercise, we will begin to put some of those pieces together. We will look at the distributions of the open clusters and the clusters associated with nebulosity. We will compare them with the position of the Milky Way and then try to understand how conditions in a galaxy could be responsible for some of these celestial objects.

Procedure
Section 30.1

Start *RedShift College Edition*, and then bring the *RedShift 3* screen to the front. In the Display menu, turn off all the items except **Deep Sky** and **Stars**. Change the zoom to **0.26**, and press the <Enter> key on the computer keyboard to shrink the sky so that it all fits on the screen.

Click on the small arrow to the right of the Aim box, choose **Deep Sky**, and select **Lagoon Nebula**. This nebula is in the Milky Way, in the constellation of Sagittarius, almost directly between us and the center of the galaxy. When you are looking at the Lagoon Nebula, you are looking toward the center of the galaxy. The sky will shift so that you can see the main part of the Milky Way.

From the Control menu, choose **Base Plane** and select the **Eq. Earth** option. This setting will rotate the sky to show how it looks from the

30

Earth's equator. Northern Hemisphere observers have the same view when the summer Milky Way is high in the sky.

Click on the Filters button, and select **Deep Sky** to open the Deep Sky filters window. In the Deep Sky filters window, de-select the **Open** and **Globular** buttons in the Clusters section by clicking on them, leaving the **With Nebulosity** clusters selected. Then click on the small arrow box to the right of the Clusters title, select **Icons**, and click **OK**. Now the display will show all the positions of star clusters associated with nebulosity. Answer the following questions.

Q1   Is there any symmetry or pattern in the arrangement of the positions of the clusters associated with nebulosity? If so, what does it look like?

Q2   Click on several different clusters in the area where the clusters seem to be centered. In the information box for each cluster you click on you will find information about the position of the cluster in the sky. What is the average declination of that area?

Q3   Is this north of the celestial equator or south?

Section 30.2

Again, click on the Filters button, and select **Deep Sky** to open the Deep Sky filters window. Click on the **Open** button in the Clusters section to select it, and then click **OK**. Now all the open star clusters will be added to the display. Answer the next question.

Q4   How do the positions of the open star clusters relate to the positions of the star clusters with nebulosity?

Section 30.3

From the Information menu, select **Photo Gallery**. In the Contents frame, click on **The Galaxy**, and select **The Milky Way**. Scroll down through the images in the Milky Way photo list to the one titled **[1093] Fish-eye lens view of the Milky Way**. Click on the **Magnify** icon to enlarge the image, and answer the next questions.

Q5   Ignoring the bright stars in this all-sky image, describe any feature that resembles the pattern created by the open clusters and the clusters with nebulosity. (The angle of this image is reversed because it was taken in the Southern Hemisphere.)

Q6 The hazy band that extends diagonally across this picture from lower left to upper right is the Milky Way. Note the part of the picture where the Milky Way is brightest, and compare its position with the area you chose as the center of the clusters in Q2.

Section 30.4

Click the Thumbnail icon to go back to the Photo gallery's Milky Way list, and scroll back to the image titled **[0830] The center of the Galaxy in the infrared**. Enlarge this picture, and compare it with the previous picture and with the display of the positions of the open clusters. Now answer the following questions.

Q7 In this image, we are seeing areas of the sky that are relatively hot. How is this photo of the hot areas of the sky similar to the positions of the clusters and nebulosity?

Q8 Again compare the center of this image with the center of the Milky Way photograph and the center of your sky display. What similarities do you see?

Section 30.5

In the Contents frame, click on **The extragalactic universe**, and select the section titled **Spiral galaxies**. Scroll about halfway down through the Spiral galaxies photo list to the image titled **[0523] NGC 3628, an edge-on spiral galaxy in Leo**. Enlarge the photo and compare it with the photos of the Milky Way and with your sky display of the open clusters.

Q9 Compare the Milky Way photos (and your sky display of the open clusters) with the photo of NGC 3628. List any similarities and any differences.

Similarities   Differences

Section 30.6

Return to the Spiral galaxies list, and scroll down to the image titled **[0996] NGC 4565, type Sb spiral galaxy**. Enlarge this picture, and compare it with the pictures of the Milky Way and with your sky display of the open clusters. Now, answer the next question.

30

Q10   Compare the Milky Way photos (and your sky display of the open clusters) with the photo of NGC 4565. List any similarities and any differences below.

           Similarities   Differences

Section 30.7

Return to the Spiral galaxies photo list and scroll nearly to the bottom to the image titled **[3078] Part of the spiral galaxy M100 (HST)**. Enlarge the picture, and answer the following questions.

Q11   Study the image of M100, and note the colors. Remembering that open clusters are made mostly of blue stars, where do most of the blue stars appear in this picture, in the core of the galaxy or in the spiral arms?

Q12   Bright nebulae are usually made of hydrogen and thus are pink. Where in this picture do you find small hazy pink blotches, the core or the spiral arms?

Q13   The plot of the open clusters and nebulosity on our sky display matches the photos of the Milky Way almost perfectly. Both look very much like the photos of other spiral galaxies seen edge-on. And we see that in other galaxies, the blue stars of open clusters, and the glowing hydrogen nebulae are all located in the spiral arms, just as they appear to be in our own Milky Way. All these observations would lead us to suppose that perhaps the Milky Way is a spiral galaxy.

If the Milky Way is a spiral galaxy that we are seeing on edge because we are in the middle of it, suggest a way in which we might actually determine the structure of our galaxy's spiral arms. (*Hint:* Suppose that we could measure distances to any star cluster in the Milky Way accurately.)

Q14   It appears that intense and rapid episodes of star formation along with massive clouds of gas and dust must be common in the arms of spiral galaxies. Such high-density star formation must be triggered by some compressing effect. Describe how such a compressing effect might be created in the arms of a spiral galaxy. (*Hint:* Consider the galaxy's rotation.)

Q15 The kinds of stars found in the spiral arms of galaxies are Population I stars, which have higher abundances of heavy elements created primarily in supernova explosions. Considering the sizes of the stars in open clusters, would the spiral arms be likely locations for supernovae? Explain your answer.

Discussion

As astronomers plotted the positions of the areas of young stars, nebulae, and brilliant open clusters, they realized quickly that these positions corresponded very closely with the path of the Milky Way through the sky. When we began to take true-color photographs of other spiral galaxies, we realized that these galaxies possess the same kinds of objects and that these objects are located almost exclusively in the spiral arms in these other galaxies. We put two and two together and realized that our Milky Way must also be a spiral galaxy, and we must be somewhere in the middle of it. We have since been able to construct a fairly accurate picture of the spiral arms on our side of the galaxy, but the arms behind the galactic disk are hidden from our view.

We have been able to see most of the spiral arm features observed in our Milky Way in these other spiral galaxies. But there is one feature of the spiral galaxies as seen from the edge that we have not yet explored in our own Milky Way—the galactic bulge in the center. It is very noticeable in the edge-on photos of some galaxies in particular. In our next exercise, we will see if we can detect evidences of such a spiral bulge in our own galaxy as well.

Conclusion

In a couple of paragraphs, summarize what you have learned about the spiral arms of our Milky Way. What kinds of observable features mark the spiral arms? What other evidence do we have that these features in the Milky Way are indicators of spiral structure? What conditions in the spiral arms would be likely to produce the kinds of stars and clusters we see?

Notes

30

The Milky Way:
The Galactic Bulge

Introduction and Purpose

In the previous exercise, we saw that the open clusters and regions of nebulosity we can see easily from Earth are aligned almost perfectly with the Milky Way. We also saw that the shape and structure of the Milky Way is very similar to that of other galaxies we see edge-on. Finally, we saw that open clusters and nebulae occur in other galaxies in the spiral arms, and concluded that what we see as the Milky Way is, in reality, the main disc of our own spiral galaxy consisting of several spiral arms all aligned with our line of sight.

However, in the edge-on images of other galaxies, we also noted a central bulge, sort of like the saucer-section of Star Trek's *Enterprise*. Can we find a similar structure in our own Milky Way? This is the challenge of today's exercise.

Procedure
Section 31.1

Start *RedShift College Edition*. If the Contents frame is not open, click on the **Contents** button in the navigation bar to open it. Click on the **Science of Astronomy** tab, and select **Photos**. In the Contents frame, click on the heading **The Extragalactic Universe**, and select **Spiral galaxies**. Scroll down through the Spiral galaxies photo list to the photo titled **[0988] The Sombrero galaxy, NGC 4594**. Enlarge this picture. (*Note:* the large bluish halo surrounding the galaxy which makes it look like a flying saucer is an artifact of image processing and is not part of the galaxy!)

Q1 In this picture, you can see the linear shape of the edge of the galaxy. Running all the way along this linear edge is a dark streak. What do you think might be causing it?

Q2 Look closely at the galaxy's disc, just above the dark streak and just inside the right edge of the galaxy. Notice the faint circular features that are just a little brighter than the rest of the galaxy. These are spiral arms. How many spiral arms can you count in this picture?

Q3   Now notice the very bright center of the Sombrero galaxy. What shape is it?

Q4   Are there dark streaks running parallel to the linear axis of the Milky Way like those in the Sombrero Galaxy? (To find out, in the Contents frame, choose **The Milky Way** from **The Galaxy** section. Now look at the image **[0829] The center of the Galaxy in visible light**.)

Section 31.2

Again, select **Spiral galaxies** in the Contents frame. Scroll through the Spiral galaxies photo list to the image titled **[0020] A spiral galaxy, NGC 2997**. Enlarge this picture, and compare this galaxy with the Sombrero Galaxy. Now answer the following questions.

Q5   Look at the spiral arms in NGC 2997. What two colors predominate in them?

Q6   On the basis of your discoveries in the last exercise on the Milky Way's spiral arms, what two kinds of objects do each of these two colors represent?

Q7   Now look at the center of the galaxy. What color predominates there?

Q8   If a blue color is representative of hot young stars as we saw in the last exercise, what kind of stars would the color seen in Q7 represent?

Q9   What shape is the central region with the color mentioned in Q7?

Q10 Look closely for dark streaks in this picture reminiscent of the dark streak running along the edge of the Sombrero galaxy and in our own Milky Way. In what part of the galaxy do you find these streaks, the center or the spiral arms?

Section 31.3

Now bring the *RedShift 3* screen to the front. In the Display menu, turn off all the items except **Deep Sky** and **Stars**. Change the zoom to **0.5**, and press the <Enter> key on your computer keyboard to shrink the sky. From the Control menu, choose **Base Plane** and select **Galactic plane**. The view will rotate so that the Milky Way runs horizontally across the screen.

From the Information menu, choose **Find** to open the Find window. Click on the small arrow to the right of the **Find:** box, select **Deep Sky**, and click on **Messier numbers**. Scroll through the Messier list, double-click on **M8**, and click **OK**. M8 is a bright nebula and open cluster in the constellation of Sagittarius in the direction of the center of the Milky Way. The view will change again to center the screen near the center of the Milky Way.

Click on the Filters button, and select **Deep Sky** to open the Deep Sky filters window. In the Clusters section, de-select all of the boxes except **Open**. Then click on the small arrow to the right of the Clusters title, select **Icons** from the drop-down menu, and click **OK**. Now the display will show the positions of the open clusters in the sky. As we saw in the last exercise, this line defines the disc and spiral arms of the Milky Way. Place a piece of transparent tape across the face of your monitor to create a line right down the middle of the Milky Way.

Section 31.4

Click on the Filters button once again, and again select **Deep Sky**. In the Clusters section, deselect **Open** clusters, and click on the **Globular** button to select it. Then click **OK**. Now the display will show the positions of the globular clusters in the sky. Answer the following questions.

Q11 Are there a lot of globular clusters exactly on the line running down the center of the Milky Way, or does this narrow region seem to be fairly empty?

Q12 Now change the Zoom to **0.26**, and press the keyboard <Enter> key so that you can see nearly all the globular clusters. The center of the screen is close to the center of the Milky Way; do the globular clusters seem to be centered near there as well?

Q13 Are the globular clusters denser near the center of the screen or at some other place?

Q14 Looking at the overall distribution of the globular clusters (you can ignore a few stragglers), would you say the globular clusters had
A) No symmetrical shape?
B) A roughly spherical symmetry?
C) A roughly linear symmetry?

Q15 Notice the very dense, roughly circular concentration of globular clusters near the center of the screen. It marks the outline of our Milky Way galaxy's galactic bulge.

31

Now open the Deep Sky filters window again, and turn the **Open** clusters back on. With a ruler, measure the width of the screen and the thickness of the line of open clusters, and use the ruler to sketch the shape of our galaxy's disc (to scale) in the space below.

Now, measure the diameter of the galactic bulge as indicated by the central concentration of globular clusters (ignoring the outlying ones). Using the same scale, add the Milky Way's galactic bulge to your sketch.

Q16 Compare your sketch with the image of the Sombrero Galaxy as to
(1) Overall shape

(2) Size of the galactic bulge

Q17 Recall from the exercise Globular Clusters that the globular clusters are mostly composed of older, yellow stars. If the globular clusters mark out the central bulge of a galaxy, how does this tie in with the colors you saw in the central region of NGC 2997 (the face-on spiral in Section 31.2)?

Section 31.5

Bring the *RedShift College Edition* screen back to the front. In the Spiral galaxy photo list, scroll down to the image titled **[0996] NGC 4565, type Sb spiral galaxy**. Enlarge this picture, and answer the following question.

Q18 We are looking at NGC 4565 from exactly on-edge, much as we see our own Milky Way. *Relative to the diameter of its disc*, is the spherical bulge in NGC 4565 larger than that of the Sombrero Galaxy or smaller? (Use a ruler to measure the length of the disk and the diameter of the bulge, and divide them to get a ratio. Then do the same for the Sombrero Galaxy, and compare the two ratios.)

Section 31.6

Click on the **Go** button in the navigation bar, select **Science of Astronomy**, **Galaxies**, and **The Milky Way**, and click on page **17**. Read and watch through page **27** to discover how all these facts tie together. You will also manipulate a three-dimensional model of our galaxy showing the relative distributions of star populations, nebulae, and star clusters.

Section 31.7

Now insert the Extras CD in the computer's CD-ROM drive. When the *RedShift College Edition* startup screen appears, click **Start**. Then, when the Extras title screen opens, click on **Story of the Universe**, and choose the section titled **The Milky Way Galaxy** for a wonderful recap of what we have learned about the Milky Way. Watch and listen to the presentation, and then answer the following questions.

Q19 How many light years across is the Milky Way's disc?

Q20 How many light years thick is the Milky Way's disc?

Q21 How many light years across is the Milky Way's bulge?

Q22 What kind of object makes up most of the Milky Way's halo?

Q23 How many light years in diameter is the Milky Way's halo?

Q24 How many spiral arms do we think the Milky Way has?

Q25 How many light years across is the core of the Milky Way?

Q26 What do we think might reside at the center of the Milky Way?

Q27 How long does it take the Sun to orbit the Milky Way?

Q28 How fast does the Sun travel around the Milky Way?

Q29 What makes us think the Milky Way has an invisible corona of dark matter surrounding it?

Discussion

As we look at color photographs of other spiral galaxies, we see an interesting trend. The spiral arms are invariably bluish in color, while the central regions are usually fairly round and are yellowish in color. We will take a closer look at this phenomenon in the next exercise.

31

Today, however, we saw that the round center of most spiral galaxies corresponds nicely to the distribution of globular clusters in our own Milky Way. While we cannot see our own galaxy's shape from the outside, we can see the distribution of our globular clusters, and these indicate that, just like the other spiral galaxies, our Milky Way has a central, spherical bulge.

The case for the Milky Way's galactic bulge is made even better when you take into consideration the color of the centers of most spiral galaxies. Most of the centers of such galaxies are yellow and thus must consist of older, cooler stars (as opposed to the hot, young stars in the open clusters that make up the arms of these spiral galaxies). Since we have already discovered that globular clusters are also made mostly of older, yellow stars, this finding anchors even more firmly our supposition that the globular clusters are the indicators of the Milky Way's own galactic bulge.

Finally, we saw that the larger distribution of globular clusters far from the center of the galaxy is also symmetric. This distribution is called the galactic halo. It stems from an earlier era in the formation of a spiral galaxy before the galaxy had begun to acquire the rotation that would eventually flatten it into a saucer-shaped disc. We will examine the rotation of various kinds of galaxies in the next exercise.

Conclusion

On a separate sheet of paper, summarize the overall structure of our Milky Way galaxy. Mention as many features as you can that indicate a spiral arm structure, any features indicative of a flattened, circular disc, and any features that might delineate a central bulge and a galactic halo. Finally, try to answer this question: "Are there any other measurements that could definitively *prove* that the Milky Way is a rotating spiral galaxy?"

Name: _____

Course: _____

Section: _____

Islands in the Sky

Introduction and Purpose

In the past two exercises, we have tried to decipher the structure of our own Milky Way galaxy. In doing so, we have compared various features of our Milky Way with photos of other spiral galaxies. Now that we have some idea of what the features of a spiral galaxy are, we would like to summarize those features and compare the features of spiral galaxies with those of other galaxies we find scattered about our universe.

Procedure
Section 32.1

Start *RedShift College Edition*. If the Contents frame is not open, click on the **Contents** button in the navigation bar to open it. Now click on the **Science of Astronomy** selection tab, and select **Photos**. In the Contents frame, click on the heading **The Extragalactic Universe**, and select **Spiral galaxies**. Scroll down through the images to the one titled **[0213] A spiral galaxy, M83 (NGC 5236)**. Enlarge this picture, and answer the following questions.

Q1 How many spiral arms can you count in this photo?

Q2 Describe as many of the features of spiral arms as you can find in this photo.

Q3 What is the predominant color of the spiral arms?

Q4 What causes this color?

Q5 What is the predominant color of the center of the galaxy?

Q6 What causes this color?

Section 32.2

Click the Thumbnail icon to go back to the Spiral galaxy photo list, and scroll down to the image titled **[0009] A spiral galaxy seen almost edge-on, NGC 253**. Enlarge this picture, and compare it with the previous picture. Now answer the following questions.

Q7   How many spiral arms can you count in NGC 253?

Q8   Is this galaxy tipped more or less than the previous galaxy was?

Q9   Are there any features seen in the other spiral galaxy that are missing from this photo?

Q10   Summarize four main features of spiral galaxies.
   (1)
   (2)
   (3)
   (4)

Q11   Think about what happens to the overall shape and size of a ball of pizza dough when an experienced pizza chef spins it and tosses it up into the air. On the basis of this observation, what might you guess about the speed of rotation of spiral galaxies, considering their shapes and dimensions?

Section 32.3

Now bring the *RedShift 3* screen to the front, and open your personal settings file. In the Display menu, deselect all items except **Stars**, **Constellations**, and **Deep Sky**. Click on the **Filters** button and select **Deep Sky** to open the Deep Sky filters window. Click on the small arrow to the right of the Galaxies title, select **Icons** from the drop-down menu, and click **OK**. This will set the sky to display all the galaxies as icons.

From the Information menu, select **Find** to open the Find window. In the Find window, click on the small arrow to the right of the **Find:** box, select **Deep Sky**, and click on **Messier numbers.** Scroll down through the list of Messier objects, double-click on **M87**, and click **OK**. The screen should rotate to center on M87. M87, also known as Virgo A, is many times larger than our own Milky Way galaxy. It is one of the largest galaxies we know of and is at the center of an enormous collection of galaxies called the Virgo Super-cluster. (Note all the galaxy symbols around M87.) Our own Local Group of galaxies is a part of the Virgo Super-cluster.

Q12   What constellation is M87 in?

Q13   Look closely at the symbol used to plot the position of M87. This is *RedShift's* symbol for an elliptical galaxy. Sketch it here.

Section 32.4

From the Information menu, select **Photo Gallery** to open the photo gallery window. In the Contents frame, click on **The extragalactic universe**, and select **Elliptical galaxies**. In the Elliptical galaxies photo list, scroll down to the image titled **[0026] Messier 87 and its globular clusters**. Enlarge this picture, and compare it with the previous pictures. Now answer the following questions.

Q14 What is the shape of this galaxy?

Q15 What color is this galaxy?

Q16 What kinds of stars would most likely be this color, young ones or old ones?

Q17 The scale of this galaxy is difficult to comprehend. Notice the tiny bright specks that look like stars scattered throughout the galaxy. Each one of these tiny specks is actually a massive globular cluster. Do these globular clusters seem to have
  (A) Roughly spherical symmetry (like the globular clusters in our Milky Way)?
  (B) No apparent symmetry?
  (C) Roughly linear symmetry?

Q18 Using a magnifying glass, do a quick count of the globular clusters along one quarter of one edge of the galaxy. Estimate the total number of globular clusters in M87 by multiplying your count by 4.

Q19 There are two other fairly bright galaxies in this photograph. On the basis of their shapes, would you guess they were
  (A) Spiral galaxies?
  (B) Elliptical galaxies?
  (C) Irregular galaxies?

Q20 As in Q11, remembering what happens to the shape of a ball of pizza dough as it is whirled and tossed into the air, what would you guess about the net speed of rotation of an elliptical galaxy like M87?

Q21 Consider the factor mentioned in the previous question, and apply it to the two smaller galaxies (mentioned in Q19) in the M87 photo. Would you guess the net speed of rotation of these galaxies was greater than, less than, or the same as that of M87?

Section 32.5

Close the Photo Gallery window. From the Information menu, choose **Find** to open the Find window. The list of Messier objects should still be displayed. Scroll back up a little, double-click on **M82**, and click **OK**. The sky will rotate to center another interesting galaxy.

Q22 What constellation is this galaxy in?

Q23 M82 is an irregular galaxy. Sketch the shape of *RedShift's* symbol for this type of galaxy.

Q24 There is another galaxy's symbol right next to M82. Click on it. What is the Messier number M of this galaxy?

Q25 What type of galaxy is it?

Q26 Sketch the shape of *RedShift's* symbol for this type of galaxy.

M81 and M82 are one of the most beautiful pairs of galaxies in the sky. With a low-power view, a modest amateur telescope can see both of them together in the same field.

Section 32.6

Now let's look at a photo of M82. Open the Photo gallery from the Information menu. Select **Irregular galaxies** from **The extragalactic universe** section. In the Irregular galaxies photo list, scroll down to the image titled **[0022] Optical image of M82 (NGC 3034)**. Enlarge this picture, and compare it with the previous pictures. Now answer the following questions.

Q27 In terms of its overall shape, what kind of galaxy would you think this might be (if you didn't know it was irregular)?

Q28 Look at the dark streaks in this galaxy. In which direction are they going with respect to the long axis of the galaxy?

Q29 You have seen galaxies with such dark streaks before. In which direction would these streaks normally run? (Look at the photos of the Sombrero Galaxy or NGC 4565 in the Spiral Galaxies section for comparison.)

32

Q30 Return to the Photo list, and read the description for this image. What might be creating the dust in these irregular lanes?

Section 32.7

Now select **Spiral galaxies** from **The extragalactic universe** section. Scroll down to the image titled **[1025] True-color image of NGC 4535, a type S(b)c spiral galaxy**. This is a barred spiral galaxy in the constellation Virgo. Enlarge the photo, and answer the following questions.

Q31 What colors are here that are typical of a spiral galaxy, and in which parts of the galaxy are they found?

Q32 What makes this galaxy different from the other spiral galaxies you have looked at? (Barred spiral galaxies are an important subtype of spiral galaxy.)

Section 32.8

In the data table below, compare and contrast the three types of galaxies we have looked at today. (If you don't have information about a particular characteristic, write **N/A** in that space.)

Data Table

Comparison of Galaxy Characteristics			
Characteristic	Spiral galaxy	Elliptical galaxy	Irregular galaxy
Overall shape			
Color(s)			
Age of the stars			
Dust lanes			
Gaseous nebulae			
Rotation			

Section 32.9

Now insert the Extras CD into the computer's CD-ROM drive. When the *RedShift College Edition* startup screen appears, click **Start**. Then, when the Extras title screen opens, click on **Movie Gallery**, scroll down to the movie titled **Measuring a galaxy's rotation**, and click on the Film icon.

303

When the movie window opens, click on the Play button, watch the movie, and then answer the following questions.

Q33 What difference do we find between one side of a galaxy and the other when it is rotating?

32

Q34 If the spectra from one part of a galaxy is blue-shifted, what does that tell us about the stars in that part of the galaxy?

Discussion

Galaxies come in three distinct types—spiral, elliptical, and irregular—with subtypes within each of these types. Spiral galaxies are the most interesting in structure and have a great variety of stars and nebulae. They have areas of high star density along the spiral arms interspersed with areas of very low star density between the arms. They also have a higher rate of rotation than most other galaxies. The brightest galaxies are usually spirals.

Elliptical galaxies rotate more slowly and are apparently formed from fairly uniform spherical clouds of matter. This uniformity allowed elliptical galaxies to break into lots of uniformly spaced stars and globular clusters. Because the distribution was so uniform, almost all the stars are moderate in size, and almost all are of similar age.

As with the stars in the Milky Way, the brightest galaxies are not necessarily the most common. There seem to be a great deal more elliptical galaxies than any other type, even though most of them are small and dim.

Irregular galaxies are difficult to characterize other than to say that they don't fit either of the other two categories. As we examine the irregular galaxies, we find that many, if not most, are (or once were) taking part in an interaction with another galaxy. We will look at some of these interacting galaxies in the next exercise.

Conclusion

On a separate sheet of paper, summarize what you have learned about galaxies. Describe the three major types of galaxies, and list their notable characteristics, their similarities, and their differences. Finally, suggest some possible reasons for the shapes and characteristics of each of the three types.

Optional Observing Project

Do the observation exercise "Galaxies" in Appendix C.

Name: _____

Course: _____

Section: _____

Cosmic Collisions

Introduction and Purpose

In the previous exercise, we mentioned that irregular galaxies often seemed to be interacting with another galaxy and that this interaction may be what has distorted an otherwise normal galaxy into an irregular one. In today's exercise, we will look at several examples of interacting galaxies and see evidence of just such distortion.

Procedure

Section 33.1

Start *RedShift College Edition*, bring the *RedShift 3* screen to the front, and open your personal settings file. In the Display menu, turn off everything except **Stars**, **Deep Sky**, and **Constellations**. Click on the Filters button, and select **Constellations** to open the Constellations filters window. Click on the **Boundaries** box to turn the constellation boundary lines on, and then click **OK**. Now set the Zoom to **50**, and press the keyboard <Enter> key.

Select **Find** from the Information menu to open the Find window. In the **Find:** box, select **Deep Sky** and **Proper names**. Scroll to the bottom of the list, double-click on **Whirlpool Galaxy**, and click **OK**. The sky will rotate to center on the Whirlpool Galaxy. The Whirlpool Galaxy, M51, is one of the most famous interacting galaxies in the sky. It is bright enough to be seen easily in most amateur telescopes.

After you have taken a quick look at this interesting galaxy, change the Zoom back to **1** so you can see the constellation names and outlines.

Q1   What constellation is this galaxy in? (You may have to move the Information box to see the constellation name.)

Q2   What constellation is it *almost* in?

Section 33.2

Now select **Photo Gallery** from the Information menu to open the Photo Gallery window. In the Photo Gallery contents frame, click on **The extragalactic universe**, and choose **Spiral galaxies**. In the photo list that opens, scroll nearly to the bottom to find the photo titled **[1157]**

Whirlpool Galaxy in Canes Venatici. Enlarge the picture, and answer the following questions.

Q3   What type of galaxy is the large galaxy in this interacting pair?

Q4   These two galaxies could actually be very far away from one another and just appear to be close together. Look carefully at the picture, paying special attention to the area between the two galaxies. What evidence can you find indicating that these two galaxies are actually close and are truly affecting one another?

Q5   Look carefully at the smaller galaxy. If this picture were zoomed in so that you could only see the smaller galaxy, what type of galaxy would you classify it as, and why?

Q6   Do you think the small galaxy is moving with respect to the large one? If so, in which direction do you think it is going—up, down, right, or left?

Section 33.3

In the Contents frame, select **Interacting galaxies** from the **Peculiar and active galaxies** subheading of **The extragalactic universe**. Now scroll down through the images to the one titled **[0216] NGC 4038-9, interacting galaxies of 'The Antennae'**. Enlarge this picture, and answer the following questions.

Q7   What kind of galaxies do you think these might have been before their collision? Why?

Q8   In the area where the two galaxies seem to be colliding, what unusual color do you note?

Q9   If these bright reddish-pink spots were in our own galaxy, what would they be?

Q10 In a spiral galaxy, what other kind of object is often associated with the objects mentioned in Q9?

Q11 List any evidence you see of the objects mentioned in the previous question.

Q12 The abovementioned objects (Q9 and Q10) are usually found in the arms of spiral galaxies. How might the conditions in the area where two galaxies are colliding be similar to the conditions in a galaxy's spiral arm? (What factors cause the objects in Q9 and Q10 in a spiral galaxy?)

Section 33.4

In the Contents frame, select **Peculiar and active galaxies** from the **Peculiar and active galaxies** subheading of **The extragalactic universe** section. Scroll to the bottom to the image titled **[3084] The 'Cartwheel' galaxy (HST)**. Enlarge the image, and answer the following questions.

Q13 There are three galaxies in the right-hand picture. Which one (or ones) of these three looks as though it had been hit?

Q14 Which of the three do you think did the hitting? Why?

Q15 The large, circular galaxy is surrounded by a ring of blue. From what you know about star colors, what kinds of stars would make that color? (The close-up at the upper left is a magnified section of the ring.)

Q16 By looking at the central part of this ring-shaped galaxy, what kind of galaxy do you think it probably was before the collision? Why?

Q17 In which direction do you think the small blue galaxy is moving? Why?

Q18 The lower left-hand picture is a magnified view of the central part of the big ring-shaped galaxy. In it you will see some tiny, blue, fuzzy spots. Remembering the scale of this photo (this is an entire galaxy like our Milky Way), what do you think they are?

Section 33.5

Summarize your findings for the three interacting galaxy pairs by filling in the data table below. Record what kind of galaxy you think each was *before* the interaction. Then state what kind of collision you think occurred: **head-on, glancing blow**, or **near miss**. Indicate whether you see any evidence of new star birth in: **0, 1**, or **2** of the galaxies because of the interaction. Finally, tell whether or not you can see any evidence of a connecting string of material **linking** the two galaxies or **pointing** from one galaxy to the other, or whether you see **no** evidence at all. In each space, write the appropriate boldfaced term. You may need to go back and review the three pictures again (Sections 33.2, 33.3, and 33.4).

33

Data Table

Interacting Galaxies					
Galaxy	Galaxy 1 Type	Galaxy 2 Type	Collision Type	Star Birth?	Connecting Matter?
Whirlpool Galaxy					
Antennae Galaxies					
Cartwheel Galaxy					

Section 33.6

Click on the **Go** button in the navigation bar, select **Science of Astronomy, Galaxies**, and **Cosmic Collisions**, and click on page **1**. Read and watch through page **7** to see some other examples of cosmic collisions.

Q19 In each example we have seen in this exercise, at least one of the interacting galaxies has been misshapen by the interaction. Given this fact, if you were to discover a new irregular or peculiar galaxy, what is one of the first things you should look for?

Section 33.7

Now insert the Extras CD in the computer's CD-ROM drive. When the *RedShift College Edition* startup screen appears, click **Start**. Then, when the Extras title screen opens, click on **The Story of the Universe**, and select the segment titled **From Big Bang to Galaxies**. Once the segment has opened, click on the last part, **Merger of galaxies**, and watch and listen to the multimedia presentation on galactic collisions.

Note those that produce results like the Whirlpool Galaxy, the Antennae Galaxies, and the barred spiral galaxies we saw in the last exercise. When the presentation is finished, answer the following questions.

In these simulations, did you see galactic mergers that resulted in galaxies that looked like:

Q20 the Whirlpool Galaxy?

Q21 the Antennae Galaxies?

Q22 a barred spiral galaxy?

Q23 After watching these simulations, what can you conclude about the possibilities resulting from various galactic collisions or near misses?

Section 33.8

Now click on the Movie Gallery button at the bottom of the window. When the Movie Gallery has loaded, click on the small **Film** icon for the movie **Merger of galaxies (1)**, and watch the movie.

Q24 Which of the three interacting galaxy pairs you examined today looks like the result of this computer simulation?

Discussion

In this exercise, we have seen just a few of the many examples of galaxies that have come close enough to one another for their mutual gravity to affect their shapes. Some of them were not affected much at all, but others were completely transformed into bizarre objects. If we did not understand the effects of gravity or saw one of the galaxies long after the other galaxy had "run from the scene," we might be very confused by some of the resulting shapes.

One of the lessons we can learn from galactic astronomy is that with the laws of our universe the way they are, given the right circumstances and conditions, anything is possible. And the corollary to that lesson is just as exciting: If we look long enough at a large enough number of galaxies, sooner or later we will run across just about every one of those possible combinations.

Conclusion

Summarize the possible effects of interactions between galaxies. Discuss the changes in shapes and in types of stars and other objects that can occur, as well as the kinds of collisions that can occur. Remark on ways in which we can confirm that two or more galaxies are really interacting and are not just superimposed along our line of sight. Finally, indicate which of the basic galaxy types is related to galactic interactions.

33

Name: _____

Course: _____

Section: _____

Galaxy Clusters

Introduction and Purpose

In the previous exercise, we examined interacting galaxies. Gravity is by nature a very weak force for ordinary amounts of mass. But if you get enough mass together, its gravitational influence can be felt over extreme distances. A spiral galaxy like our Milky Way has a mass of around 400 billion times that of our solar system. This much mass creates a gravitational force that can be felt tens of millions of light years away.

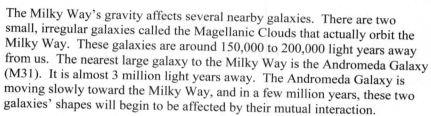

The Milky Way's gravity affects several nearby galaxies. There are two small, irregular galaxies called the Magellanic Clouds that actually orbit the Milky Way. These galaxies are around 150,000 to 200,000 light years away from us. The nearest large galaxy to the Milky Way is the Andromeda Galaxy (M31). It is almost 3 million light years away. The Andromeda Galaxy is moving slowly toward the Milky Way, and in a few million years, these two galaxies' shapes will begin to be affected by their mutual interaction.

In addition to the Andromeda Galaxy, there are several dozen other galaxies that are all bound together by their mutual gravitation. We call this collection of galaxies The Local Group. As we look out into space, we see other groups of galaxies clumped together in different parts of the sky. As we have been able to plot galaxies' distances more accurately, we have begun to discover even larger, more loosely knit conglomerations of galaxies whose motions seem to be affected by the overall gravity of many galaxy clusters spread over huge reaches of the universe.

However, since the distances to these galaxies are only approximately known, we must always ask the question: Are all the galaxies that appear to be close together in the sky actually near one another or just accidentally aligned? In today's exercise, we will examine a few of these clusters of galaxies and learn some of the ways in which this question is answered.

Procedure
Section 34.1

Start *RedShift College Edition*. If the Contents frame is not open, click on the **Contents** button in the navigation bar to open it. Click on the **Science of Astronomy** tab, and select **Photos**. In the Contents frame, click on **The Extragalactic Universe** section, and select **Galaxy groups**. In the

Galaxy groups photo list, scroll down to the image titled **[1059] Stephan's Quintet**. This is a very compact cluster of galaxies containing four (maybe five) galaxies. Enlarge this picture and answer questions Q1 through Q8.

Q1 How many galaxies can you count in this picture? (There is one more galaxy in this group that is just off the edge of the photo at the lower right.)

Q2 List as much evidence as you can find that these galaxies are truly near one another. (Find as much evidence as you can of interaction between the galaxies.)

Q3 Three of these galaxies are the same color—yellowish. What kind of galaxy would you expect to be yellowish, a spiral or an elliptical?

Q4 Do the shapes of these three yellowish galaxies match their color? Do they look like elliptical galaxies, or do they show evidence of spiral arms?

Q5 Does the shape of the fourth galaxy match its color?

Q6 The fourth galaxy (the bottom one) has a much lower redshift than the other three. According to the Hubble Relation, the lower a galaxy's redshift, the closer the galaxy is to Earth. If so, then the fourth galaxy is actually a great deal closer to us than the other three and is not actually a part of the cluster or a part of any interaction with the others. So here is the debate: Is the bottom galaxy really close to us and just accidentally aligned with the other three, or are they all part of the same interacting cluster?

Look carefully at the top three galaxies and at the lower one. Can you find any evidence that the lower galaxy is at the same distance as the other three and is actually interacting with them?

Q7 The more dust and gas a star's light passes through, the redder the light gets (like the sun's light at sunset). So the farther away in space an object is, the redder it appears because of filtering by intergalactic dust and gas. With this in mind (along with the discussion already presented), refer to the debate stated in the previous question. Again looking carefully at the four galaxies in this picture, give as much evidence as you can that the bottom galaxy is actually a lot closer to us than the others, and is *not* an interacting part of the cluster.

Q8 Weigh the evidence you listed in favor of the four being at the same distance (Q6) and the evidence you listed against this argument (Q7). Which arguments do *you* think are most persuasive, and why?

Section 34.2

In the Contents frame, select **Galaxy clusters** from **The extragalactic universe** section. In the Galaxy clusters photo list, scroll down to the image titled **[1170] Hercules cluster of galaxies**. Enlarge this image, and answer the following questions.

Q9 We want to estimate the number of galaxies in this picture. First, divide the picture into four equal parts, and count the number of all galaxies (anything not perfectly round) in the lower right quadrant.

Q10 To obtain an estimate of the total number of galaxies in this cluster, multiply your answer to Q9 by 4.
Total number of galaxies in the Hercules Cluster = (**Q9**) * 4
$$= (\quad) * 4$$
$$= (\quad\quad)$$

Q11 How many of the galaxies you counted in the lower right-hand quadrant of this picture (Q9) look like spiral galaxies?

Q12 What kind of galaxy do the rest appear to be?

Q13 On the basis of your answer to the previous questions, which kind of galaxy seems to be more common, spiral or elliptical?

Q14 Now take a quick glance at the picture and tell which kind of galaxy seems largest, brightest, and most obvious?

Q15 How many galaxies can you find in this picture that show clear signs of interaction?

Section 34.3

Click the Thumbnail icon to return to the Galaxy clusters photo list, and scroll down to the image titled **[3088] Gravitational lensing by the galaxy cluster Abell 2218**. Enlarge this image, and answer the following questions.

Q16 We want to estimate the number of galaxies in this picture. As before, divide the picture into four equal parts and count any object that is yellow unless it is perfectly round or very tiny. How many galaxies are in the upper right quadrant of the picture?

Q17 To obtain an estimate of the total number of galaxies in this cluster, multiply your answer to Q16 by 4.

Total number of galaxies in Abell 2218 = (*Q16*) * 4
= (   ) * 4
= (   )

Q18 On the basis of their shapes, what kind of galaxy are most of these?

Q19 Now look carefully at the bluish objects, particularly those around the center of this image. There is something strange about their shapes. What is it?

Q20 There is also something interesting about the symmetry and placement of the bluish "arcs." Imagine these arcs extended to form complete circles. What do you find where the geometric center of these strange circles would be?

Q21 Einstein's theory of general relativity predicts that a large enough concentration of mass can actually bend light like a lens. Such a gravitational lens could form a magnified image or simply distort the light coming from a galaxy much farther away. How does this theory help explain the shape and symmetry of the bluish objects?

Q22 If Einstein's theory is the correct explanation of this strange picture, what does this finding imply about the galaxies (both size and number) in the cluster Abell 2218? (Would you expect this cluster to have a lot of mass or not?)

Web Intrigues

Click the selection tab (Currently set to **Photos**), and select **Web** to open your browser window. Click in the Address box and enter the following URL address: *http://oposite.stsci.edu/pubinfo/* This is the Hubble Space Telescope page. Select the **Hubble Pictures!** link to load the HST Public Pictures page. On this page, select the link titled **1999 Releases**. Finally, select the press release titled **A Minuet of Galaxies**. This image shows a Hubble Space Telescope photograph of the galaxy cluster known as Hickson 87. (You can see a large version of this photo by selecting the **Press Release Images** link and then clicking on the **JPEG** icon.) Study the photograph carefully, thinking about what you have learned in this exercise.

Q23 Which of these galaxies are clearly interacting?

Q24 Which of these galaxies looks as though it might not belong with the rest?

Q25 Would you guess the galaxy you chose in the last question is closer than the rest or farther away? Why?

Section 34.4

Now insert the Extras CD into the computer's CD-ROM drive. When the *RedShift College Edition* startup screen appears, click **Start**. Then when the Extras title screen opens, click on **Movie Gallery**, select the segment titled **Gravitational lensing**, and click on the Film icon to open the movie window. When the movie window opens, click on the Play button, and watch this animation to see how gravity forms such effects.

Discussion

Determining the positions, sizes, and motions of clusters of galaxies is an ongoing project and will continue to be in the foreseeable future. Our ability to measure distances to most of these clusters of galaxies is not very good at the present. However, what we have begun to discover is fascinating. Just as the mass around a star clumps together to form a solar system, the stars clump together to form open and globular clusters, and the star clusters clump together to form galaxies, so galaxies themselves clump together to form galaxy clusters.

Now we have discovered galaxy super-clusters formed from the galaxy clusters, and finally, walls of super-clusters forming what appear to be bubbles in the huge emptiness of our universe.

One of the questions we are trying to answer by studying galaxy clusters is whether or not the Hubble Relation is accurate and whether we can continue using the redshift of a galaxy to determine its distance. Some galaxy clusters (like Stephan's Quintet) seem to show that redshift is not a reliable indicator of galactic distance. However, even these apparent discrepancies may have other, better explanations.

34

Conclusion

Summarize what you have learned about clusters of galaxies. Include typical sizes of the clusters and the relative proportions of different kinds of galaxies in them. Describe the evidence that shows that a galaxy truly is part of a cluster, and describe how you might tell that one particular galaxy did not belong with the rest. (Continue your conclusion on a separate sheet of paper if necessary.)

V
Appendices

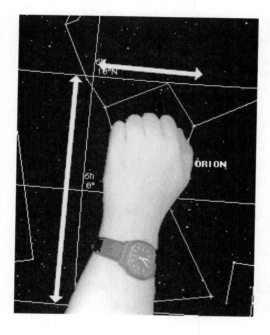

Observing With RedShift

Contents

Starting *RedShift College Edition* for the First Time

Procedure
Section 1.1
Windows:

Insert the *RedShift College Edition* CD #1 into your computer's CD-ROM drive, and close the drive door. After a few seconds, the CD-ROM drive should spin, and a small window should appear on your screen, titled *RedShift College Edition*. Click on the **START** button.

Macintosh:

Insert the *RedShift College Edition* CD #1 into your computer's CD-ROM drive, and close the drive door. After a few seconds, the CD-ROM drive should spin, and a small folder with several icons should open on the desktop. In this folder, locate the icon titled *RedShift*. (Scroll around in the folder if necessary.) Double-click on this icon to start *RedShift 3*.

Section 1.2

A small screen should now appear stating that *RedShift 3* is loading. This screen will be followed by the Brooks/Cole logo and the *RedShift 3* title screen all accompanied by a musical fanfare. If you don't hear any music, check your speakers or the volume control on your computer desktop.

Finally, the *RedShift 3* window and then the *RedShift College Edition* window will open on your screen. The *RedShift College Edition* package you have just installed consists of two separate programs—a special version of *RedShift 3* and *RedShift College Edition*. The specially modified version of *RedShift 3* contains the main features of the standard *RedShift 3* software package plus a special set of tools for making measurements on the *RedShift* sky. *RedShift College Edition* contains an interactive, narrated tutorial called "The Science of Astronomy" and functions as a hub from which you can access the World Wide Web and most of *RedShift 3 's* amazing simulations, photos, and prediction tools.

The software automatically opens the *RedShift College Edition* window on top. However, the exercises in this book will work directly with *RedShift 3* most of the time, so you will have to switch to the *RedShift 3* window each time you start a new exercise. You can switch back and forth between the two windows at any time by clicking on the **Window** menu at the very top of the screen and selecting *RedShift 3/RedShift College Edition*.

Section 1.3

If this is your first time to start *RedShift College Edition*, you will need to set some initial parameters so the software will make proper calculations for your viewing location. The Preferences window should open automatically the first time *RedShift College Edition* is started. (If it does not, select **Preferences** from the **File** Menu to open the Preferences window.)

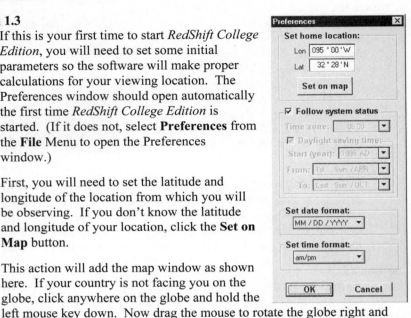

First, you will need to set the latitude and longitude of the location from which you will be observing. If you don't know the latitude and longitude of your location, click the **Set on Map** button.

This action will add the map window as shown here. If your country is not facing you on the globe, click anywhere on the globe and hold the left mouse key down. Now drag the mouse to rotate the globe right and left and up and down until your country is visible in the top part of the map window. Now click on the Magnifying glass icon just below the upper map window to put city names on the map. Position your mouse pointer as close as you can to your location and double click with the left mouse button. This should shift the map so that the yellow cross-hair is over your location.

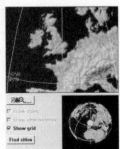

Once the latitude and longitude have been set, you need to check the **date format** and the **time format** and set them as shown in the illustration above. Finally, click the **OK** button to save your preferences.

Section 1.4

Before we go any further, we need to set the parameters of the sky display with the features you will be using most often.

First, open the Display menu and click on the item labeled **Natural Sky Color** to turn the blue sky background **Off** during daylight hours. Now you should be able to see the stars on the *RedShift* screen, even during the daytime. Again from the Display menu, click on the **Milky Way** item to remove its check mark, and turn the Milky Way overlay **Off**.

Finally, open the Display menu and click on the **Constellations** item if it does not have a check mark. This will turn **On** the constellation pattern overlays.

Now check the panels shown on the screen. The following panels should be open on your screen: **Time**, **Control Time**, **Location**, and **Aim**.

If any of these panels is not open, open it by clicking on the appropriate small icon button in the **Settings** and **Activities** panels. For the moment, open the Control Panel by clicking on the small Control Panel icon ⬕ in the Activities Panel.

When the Control Panel opens, click on the number **10** in the

Step, box and type in **5** to reduce the number of degrees by which the screen moves every time a directional arrow is pressed. Click on the small Control Panel icon once more to close the Control Panel.

Now, click on the small arrow to the right of the Aim box in the Aim panel. From the menu that drops down, select **Stars**, and then click on **Polaris** as shown. The view will rotate to put Polaris, the North Star, in the center of your screen.

Section 1.5

Now we are ready to do a little customizing of your sky display. Each person has a different idea about what makes the sky easiest to use, and *RedShift* allows you to set many of the display parameters to your own liking. Click on the panel button labeled **Filters**, and from the menu that drops down, select **Guides**. In the **Local** section of the Guides filters window, click on the small **Horizon** box, and select **Line** from the drop-down menu. Click on the small **Stars** tab at the left of the screen to switch to the Stars filters. Selecting any of the tabs on the left side of this

window will change the Filters window to that particular type of object.

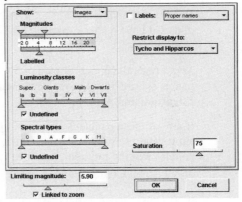

In the Stars filters window are several sections with slider controls. We will set the sky for a typical dark sky site high in the mountains or way out in the desert. Set the upper **Magnitudes** slider and the **Limiting magnitude** slider as shown. Change the entry in the data selector **Restrict display to:** so that it reads **Tycho and Hipparcos** as shown.

As you make each change, notice how the small **Show Preview** display changes. This display allows you to preview any changes you may wish to make before resetting the entire sky. When you have made all these changes, click **OK** to apply the changes to the entire sky.

Now look at the stars of the Little Dipper (Ursa Minor). You will see that all the stars are now brighter and that there are more stars. If this view seems too artificial to you, you may wish to reopen the Stars Filter window and tinker with these adjustments. The upper **Magnitudes** sliders allow you to set the brightest and dimmest stars to show. The upper left-hand slider determines the magnitude of the brightest stars displayed. Move it down to **5**, and look at the Preview display. You have just eliminated all stars brighter than 5th magnitude—those easiest to see. Restore the left-hand slider to its maximum position.

Now slide the upper right-hand slider all the way to the right. This tells the computer to display even the very faintest stars in its catalogue. You will not immediately see the effects of this setting however, because of the **Limiting magnitude** slider. The **Limiting magnitude** slider lets you determine the sensitivity of your view. The sensitivity of the typical human eye in a very dark sky is about 6th magnitude, so we suggest it be set to that value. However, if you wish to see where all the fainter stars are, drag the **Limiting magnitude** slider all the way to the right, and look at the Preview display. You should see so many stars that it will be hard to see any blank sky. For maximum usefulness, leave the upper right-hand **Magnitudes** slider set at its maximum position, and vary the sensitivity of the view with the **Limiting magnitude** slider.

There is one last option to note here: the **Linked to zoom** box. When this option is checked, the sensitivity of the view screen increases as the telescope magnification is increased, just as a real telescope would do. This allows you to see a normal sky at a zoom of **1** but still see the very faintest stars when the zoom is set to a maximum value. We will almost always leave this option checked.

When you think you understand how each control works, set the sliders to create a sky most like what you are used to looking at. We recommend that you set the upper **Magnitudes** sliders to their extreme positions and then vary the view from a city view to a dark sky view by adjusting the **Limiting magnitude** slider. Click **OK** to see the effects of your settings. You may have to reset the **Limiting magnitude** slider several times to get a sky that looks like the one you're familiar with.

Once you have the **Limiting magnitude** set as you like, again open the Stars Filter window, and turn to the **Saturation** slider. Move it to its maximum position, and click **OK**. What happens to the colors of the stars? Move it all the way to the minimum position, and click **OK**. Now what are the colors? The ability to show star color makes *RedShift* a powerful tool for identifying different types of stars, but it can be confusing because our eyes do not detect any but the brightest star colors. Try several different settings to get the **Saturation** slider set to a level you find easiest to use.

When you have set the sky display to your satisfaction, go on to Appendix B to learn how to save your settings so that you don't have to go through this process each time you start *RedShift*.

Notes

A

Saving and Opening a Personal Settings File

I. Saving Your File

After you have the display set to your personal tastes (and to the workbook specifications in Exercise 2), you will need to save a Personal Settings file. From the File menu, select **Save Settings As...** to open the **Save As** dialog box as shown.

A. Your Own Computer

If you are working on your own computer, click in the box titled **File name**, and type in a unique name (preferably eight letters or less) that you will be able to remember later. Use your initials if you like, but be sure

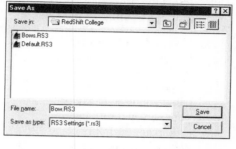

to leave the extension **.RS3**. Your completed file name should now be something like this: **Bow.RS3**. When you have entered an appropriate file name, click on the **Save** button to save your customized personal settings.

B. Classroom Computer

If you are using a classroom computer, you will need to place a blank floppy diskette in the computer's disk drive and then click on the small

arrow to the right of the *RedShift College* folder name (as shown). From the list of folders and drives that will appear, select **3½ Floppy (A:)**. (Use the vertical scroll bar if necessary to locate drive **A:**.) At this point, the floppy drive should activate and load a blank directory.

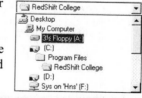

Now, click in the box titled **File name**, and type in a unique name (preferably eight letters or less) that you will be able to remember later. Use your initials if you like, but be sure to leave the extension **.RS3**. Your completed file name should now be something like this: **Bow.RS3**. Finally, click on

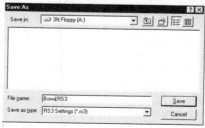

B

the **Save** button to save your customized personal settings.

II. **Opening Your File**

Whenever you use *RedShift 3*, you will usually want the sky set up in the same way as it was when you saved your personal settings file. To restore these settings without having to go through the tedious process of making each change individually, you can reload all the settings at once with the following procedure. First, click on the File menu, and select **Open Settings** to bring up the **Open** dialog box as shown.

A. **Your Own Computer**

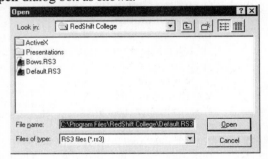

If you did not change any directories when you saved your file, the file name you used in saving your personal settings file should appear in the directory window (for example, **Bow.RS3**, as shown). Click on your file name to select it, and then click the **Open** button to restore your personal settings to the sky display.

B. **Classroom Computer**

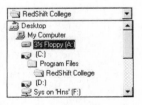

If you are using a classroom computer, you will need to place the floppy diskette (which you used to store your settings file) in the computer's disk drive and then click on the small arrow to the right of the *RedShift College* folder name. From the list of folders and drives that will appear, select **3½ Floppy (A:).** (Use the vertical scroll bar if necessary to locate drive **A:.**) At this point, the floppy drive should activate and load the diskette directory.

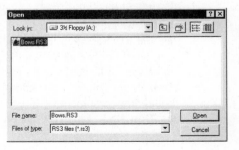

The only file that should show up in the directory window will be your personal settings file (for example, **Bow.RS3**, as shown). Click on the file name, and then click the **Open** button to restore your personal settings to the sky display.

Observing Projects

Constellations

Introduction and Purpose

As we found in the exercise "'Landmarks' in the Sky," constellations provide us with much valuable information about where to find things. As you begin exploring the sky with binoculars or telescopes, you will find that there are odd and memorable groupings of stars on smaller scales as well. The more you explore the sky, the more of these groupings you will come to recognize.

By using these distinctive patterns of stars, you can easily learn to plot a path to just about any place in the sky you might want to go. With *RedShift* along to provide a "road map," you can quickly locate identifiable "signposts" or "landmarks" to guide you, landmark to landmark, until you arrive at your destination galaxy, nebula, or star cluster. We call this process *star-hopping*, because that's just what we do—hop from star to star and pattern to pattern to find the object we wish to observe.

By learning to recognize star formations, we make the sky a familiar place, just like the neighborhood in which we grew up. If you spend enough time exploring the sky, you will get the same sense of excitement as you turn a corner with your telescope and spy an old familiar celestial landmark as you do when you come home after being gone a long time.

In today's observation exercise, one purpose is to use *RedShift* to locate some of these celestial guideposts to help you learn to star-hop. The second purpose is to give you some experience in using your built-in "sky ruler"—your hand—to use *RedShift's* star-hopping information to actually find things in the sky.

Procedure

1. The first step in doing this exercise is to sit down with *RedShift* and plan a star-hop to take you from a constellation you can recognize easily to each of the constellations listed in the data table on page 328. In the data table are a set of fall/winter constellations and a set of spring/summer ones. Choose the set appropriate for you. Once you have decided which constellations you want to observe, you will use *RedShift* to generate a star chart or charts with star-hops marked on them. (Follow the procedures found in the section "Star-Hopping" of Appendix D to generate these charts.)

C1

2. The first clear evening after you have prepared your charts, get out your binoculars, a soft pencil and eraser, and a flashlight covered with red cellophane, and head for a good, dark observing site. Spend a few minutes letting your eyes adapt to the dark and locating any constellations you can. After about 10 or 15 minutes, you are ready to begin observing. Find your jumping-off constellation and, using the red flashlight to look at your star charts, compare the scale of your printout with the real sky. This comparison will give you some sense about how far away each constellation will be and how large it will appear. You may also have to turn your printout to match up with the real sky directions.

Once you are oriented to the real sky, begin looking for your star-hop landmarks. One, by one, follow them to your destination. Once you find the stars that match the constellation you wish to observe, begin filling out the data table below. Using your hand as a "sky ruler" (see "Sky Rulers" in Appendix D), estimate the constellation's altitude and azimuth and the width and height of the constellation. Then remark on the overall brightness of its stars and their colors, mark on your star chart which of the stars of the constellation appears to be brightest, and record the date and time of your observation. Continue in this manner until you have observed four of the six constellations for your particular season.

Data Table

Constellations							
Constellation	Altitude	Azimuth	Height	Width	Remarks	Date	Time
Spring/Summer							
Auriga							
Leo							
Bootes							
Corvus							
Corona Borealis							
Hercules							
Fall/Winter							
Sagittarius							
Cygnus							
Pegasus							
Taurus							
Auriga							
Gemini							

The Moon

Introduction and Purpose

The Moon is one of the most impressive sights in the night sky; its motion through each month has affected many aspects of human life and culture. In this observation exercise, we will observe the moon as often as possible during a two-week period and plot its daily positions on a star chart to show just how the Moon moves through the sky. In doing so, we will gain a better appreciation of how far the Moon moves each night, how its phases are related to its rising and setting times, and what constellations it passes through on its monthly journey.

Procedure

1. The first step in doing this observation exercise is to prepare a star chart of the part of the sky where the Moon is tonight, showing the main constellations. This star chart will be your data table, where you will plot the daily positions of the Moon for a month. Start *RedShift College Edition*, bring the *RedShift 3* screen to the front, and open your personal settings file. Change the sky settings as shown in the table below.

Date	Today's date
Zoom	**0.5**

In the Display menu, turn off all items *except* **Moons, Stars, Constellations**, and **Guides.**

2. This exercise should be started one or two days after new moon, when the moon is a thin crescent in the evening sky just after sunset. To determine when to begin your first observation, select **Sky Diary** from the Information menu at the top of the screen. Once the Sky Diary window has opened, look through the list of events for the month (scrolling if necessary) until you find the event titled **New Moon**. See what day of the month this occurs on, and then plan your first observing session right after sunset two days after this date. Set the *RedShift* date box two days after new moon, and press the <Enter> key on the keyboard to set the sky for that date.

3. To locate the moon with respect to the stars, click on the Filters button and select Constellations to open the Constellations filters window. Click on the **Full Latin** tab, and select **Abbr. Latin** from the drop-down menu to change the constellation names to abbreviations. In the upper right-hand

section of the window, click the button titled **None** to deselect all the constellation names, and then click the button labeled **Zodiacal** to turn only the Zodiac names back on.

Now, click on the small tab at the left of the window labeled **Guides**. In the Guides filters window, click on the **Ecliptic** box to place a check mark there. In the Ecliptic section, click on the **Grid** box to remove its check mark. Finally, click on the **Ecliptic plane** box to turn the Ecliptic plane **On**. Now click on the Celestial (Ra/Dec) box to activate that section. When your Guides filters window looks like the picture shown here, click **OK** to close the window. Now you should see the abbreviations of the zodiacal constellation names and a single red line running across the sky.

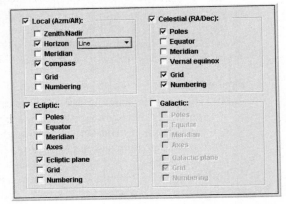

C2

4. At this point, you are almost ready to print your "Moon-plot" star chart. This is the chart you will be using for the first few days to plot the daily positions of the Moon, so look at it carefully. The red line is the ecliptic—the path the Sun follows—and is quite close to the Moon's path. It will be a rough guide to help you locate the Moon's position each night on the star chart. The constellation names and patterns may be more confusing than helpful once you start plotting the Moon positions, so you may wish to note one or two and then turn them off before you print the sky display. Once you have decided whether to keep the constellation names and patterns on the display or turn them off, choose **Print** from the File menu.

5. When the evening of your first observation date arrives, spend a few minutes letting your eyes get adapted to the dark and locating all the constellations you can. Locate the Moon, and note the stars in the area around it. Find the same stars on your printout, and sketch a tiny moon at its actual location on the printout showing the approximate phase. Now estimate the right ascension of the Moon to the nearest tenth of an hour from the grid on the printout, and record it along with the date and time in

the data table on page 332. Finally, observe the Moon through binoculars (if you have them) and sketch it more accurately in the space provided in the data table. Sketch not only its phase accurately, but include any details you note through the binoculars, including major craters, ray patterns, and maria.

6. Repeat Step 5 every other night (or as close to that interval as the weather permits) until the Moon is near the left edge of your original chart (approximately four days later). When this happens, go back to *RedShift*. Set the date for eight days after your original chart date, adjust the time to an appropriate viewing time for the new date, and center the Moon. The current Moon position should just be visible at the right edge of the new chart. If not, adjust the date until it is, and then print the new star chart.

   As before, plot the Moon positions every couple of nights on the new chart until it, too, needs to be replaced. **Continue this process until the Moon *does not rise until after midnight*.** At this point you should have two or three Moon chart printouts covering approximately fourteen to sixteen days.

7. Find a pair of successive observations that were exactly two days apart. Subtract the right ascension of the Moon on the second night from its right ascension on the first (subtract the hours and the tenths). How many hours (including tenths) of right ascension did it move in two nights?

   Now divide that number by two to get the length of time the Moon's rising is delayed each successive day.

When you are done, attach your moon-plot star chart to the back of this exercise and turn it in.

Data Table

Date	Time	RA	Sketch

C2

The Planets

Introduction and Purpose

The planets were of great importance to many ancient cultures. Their somewhat unpredictable motions through the sky led early astronomers to ascribe personalities and mystical powers to them. In this observation exercise, we will observe one of the naked-eye planets as often as possible during a one-month period and plot its daily positions on a star chart to show just how it moves through the sky. In so doing, we will gain a better appreciation of how far these planets actually move each night and how that motion varies over time.

Procedure

1. The first step in doing this exercise is to prepare a star chart of the part of the sky containing your planet showing the main constellations. This chart will be your data table, where you will plot the daily positions of your planet for a month. Start *RedShift College Edition*, bring the *RedShift 3* screen to the front, and open your personal settings file. Change the sky settings as shown in the table below.

Date	Today's date
Zoom	0.5

In the Display menu, turn off all items *except* **Sun & Planets**, **Stars**, **Constellations**, and **Guides.**

2. Next, choose a planet to observe. If you do not know what planets are visible in the evening sky, either ask your instructor or check out their positions with the **Sky Diary** function found in the Information menu. If you want to use the Sky Diary, once you have opened the Sky Diary window, click on the tab labeled **Overview-table and planet positions for 15th of month**. Any planet shown to the left of the Sun's position or within four hours of the right edge of the display is in the early evening sky. The planets to check are Venus, Mars, Jupiter, and Saturn. After choosing one, click **Close** to close the Sky Diary window.

Click on the small arrow to the right of the Aim box, select **Planets/Moons**, and click on the planet name you chose. The screen will rotate to center on this planet. Click on the planet to open its name tag,

C3

and then click on the name tag to open its information box. In the information box, click on the **Report** button. In the Report window that opens, click on the **Period** box, and select **30 days** to generate a visibility report covering the next 30 days.

When the report is finished, look at the graph. The vertical axis of the graph represents the hours of the day going from 12 noon at the bottom of the graph to midnight in the middle of the graph, back to 12 noon of the next day at the top of the graph. The horizontal axis represents the dates you chose. The main part of a typical report window is shown below.

Note the area in green. This represents the times during which the planet is below the horizon (not a good time to observe it). The area in white is daylight (also not a good time to observe a planet). The area in gray is

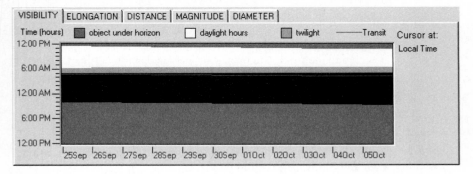

twilight (you *could* observe then, but it wouldn't be very good). The only good times to observe the planet are shown in black.

In the example shown here, you could observe this planet any time from around 10 p.m. till dawn. Generally, the best time to observe a planet is when it is highest in the sky. This point is represented by the red line on the graph—the transit point. In the example shown above, the transit occurs at around 5:30 a.m. on September 25 but shifts to around 5:00 a.m. by October 5.

If you're happy with the observing conditions presented by this planet, click the **Print** button to make a printout of this visibility report. If this planet is still too near the horizon at a reasonable time, go back to the main *RedShift* screen and try another planet. Once you have settled on a planet to observe and printed a visibility report, close the Report window and the planet's information box.

3. Click on the **Filters** button, and select Constellations to open the Constellations filters window. Click on the **Full Latin** tab, and select **Abbr. Latin** from the drop-down menu to change the constellation names to abbreviations. In the upper right-hand section of the window, click the button titled **None** to deselect all the constellation names, and then click the button labeled **Zodiacal** to turn only the Zodiac names back on.

Now, click on the small tab at the left of the window labeled **Guides**. In the Guides filters window, click on the **Ecliptic** box to place a check mark there. In the Ecliptic section, click on the **Grid** box to remove its check mark. Finally, click on the **Ecliptic plane** box to turn the Ecliptic plane **On**. Click on the **Celestial (Ra/Dec)** box to activate that section. When your Guides filters window looks like the picture shown here, click **OK** to close the window. Now you should see the abbreviations of the zodiacal constellation names and a single red line running across the sky.

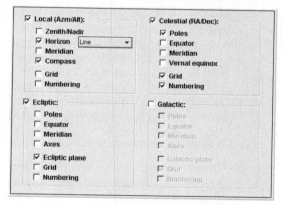

4. At this point, you are ready to print your planet star chart. This is the chart you will be using for the next month to plot the daily positions of your planet, so look at it carefully. The red line is the ecliptic—the path the Sun follows—and is quite close to the path of the planets. It will be a rough guide to help you locate your planet each night. The constellation names and patterns may be more confusing than helpful once you start plotting the planet's positions, so you may wish to note one or two constellations and then turn them off before you print the sky display. Once you have decided whether to keep the constellation names and patterns on the display or turn them off, choose **Print** from the File menu to print out your star chart. Now you are ready for your first observation.

5. When evening arrives, spend a few minutes letting your eyes adapt to the dark and locating all the constellations you can. Locate your planet and note the stars in the area around it. Find the same stars on your printout, and sketch a tiny circle at the planet's location on the printout. Now, from the grid on the printout, estimate the right ascension of the planet to the nearest 1/10 hour, and record that along with the date and time in the data table on page 337.

C3

Repeat Step 5 every other night (or as close to that interval as the weather permits) for one month.

6. Subtract the right ascension of the planet on the second night of observation from its right ascension on the first (including the tenths). Now divide the difference by the number of nights between the two observations. How much did the planet move in right ascension each night at the first of the month? (If you did not observe any change, write 0.)

7. Subtract the right ascension of the planet on the last night of observation from its right ascension on the next-to-the-last night. Again, divide the difference by the number of nights between the two observations. How much did the planet move in right ascension each night at the end of the month? (If you did not observe any change, write 0.)

   Was the planet speeding up, slowing down, or moving steadily?

8. Subtract the right ascension of the planet on the last night of observation from its right ascension on the first night. Again, divide the difference by the number of nights between the two observations. How much did the planet move in right ascension during the whole month?

   In which direction was the planet moving with respect to the stars—east or west?

When you are done, attach your planet-plot star chart to the back of this exercise and turn it in.

Data Table

The Planet						
Date	Time	RA		Date	Time	RA

C3

Notes

Star Colors

Purpose

In this exercise, we will observe the spectra of several different stars and thus identify their different temperatures.

Procedure

With a pair of binoculars, observe the following bright stars when they are near the horizon. They will be twinkling vigorously because you are looking through hundreds of miles of the Earth's atmosphere. As you watch each star, note how it alternately flashes different colors. These are the colors of its spectrum. The cooler stars will flash lots of reds and oranges and occasionally some greens, but seldom any blues or violets. The hotter stars will flash lots of violets, blues, and greens, but seldom any oranges or reds. If you are not sure where the various stars in the list are located, use the following directions to print out a star chart for finding them.

1. Start *RedShift College Edition*, bring the *RedShift 3* screen to the front and open your personal settings file. Change the sky settings as shown in the table below.

Date	Today's date
Zoom	0.5

2. In the Display menu, turn off all items *except* **Stars**, **Sun and Planets**, and **Guides.**

3. Before you actually rotate the sky to locate the desired stars, it might be useful to add the star names for the brighter stars. Click on the small button near the top of the Control Panels titled "Filters," and select "Stars" from the drop-down menu.

In the upper right part of the Stars Filters window you will see a box titled **Labels**. Click on the box to add the proper names for the brighter stars. You can adjust the number of stars with names by changing the lower **Labelled** slider in the top left part of the window. The default setting of **4** will create a pretty busy display, so you might want to drag the **Labelled** slider to the left to somewhere around **2** for this exercise. Then click **OK** to turn the labels on.

C4

Now, the bright stars should all have names. To find your stars, click on the small arrow to the right of the Aim box, and select **Stars** to open the pop-up Stars menu. Now find one of the stars you wish to observe, and click on it. This will rotate the sky to center the star you selected.

If the star you chose happens to be below the horizon from your viewing location at the time you chose, the center of the sky will appear green and you will need to choose another star (or another time of night).

4. Now you need to find out just where in the sky you are looking. If the horizon shows up at the bottom of the screen, you will see the directional indicators of N, E, S, and W. This will let you know in which general direction to look. (If you can't see the horizon at the bottom, use the down arrow key on your computer keyboard to scroll the sky until the horizon shows up.)

   Print out the chart by selecting **Print** from the File menu, and you should be ready for your observation session.

5. Observe two of these stars: Aldebaran, Antares, Arcturus or Betelgeuse. In the data table, record the following: (1) the colors you see in their spectra; (2) what you think their spectral class might be; (3) your estimate of whether they are hot, average, or cool stars; and 4) the date and time of your observation. (If you need to review spectral types, start *RedShift College Edition*. Click the **Go** button in the navigation bar, select **The Science of Astronomy**, **Stars**, and **Types of stars**, and click on page **21**. Read through this page to see the different spectral classes and their relative temperatures.)

6. Observe two of these stars: Capella, Altair, Denebola, or Procyon. As before, note which colors are present in their spectra, and classify them as to temperature and spectral class.

7. Now observe two of these stars: Sirius, Spica, Regulus, or Vega. As before, note which colors are present in their spectra, and classify them as to temperature and spectral class.

8. Finally, when you are finished, check your spectral class and temperature estimates by bringing the *RedShift 3* screen to the front. Open the Find window from the Information menu, and select **Stars** and **Proper names** in the **Find:** box. Locate each star in the list, and double-click on its name. Then click the **Information on...** link in the **List Links** box to open the star's information box. In the information box, click on the **More** button, and select the **Color & Luminosity** tab to find this information. (Alternately, you can find these stars listed in the appendices of most textbooks.) Then write the actual spectral class and temperature of each star below your estimates.

Data Table

Bright Stars				
Star	Colors	Spectral Class	Temp	Date/Time
Aldebaran				
Antares				
Arcturus				
Betelgeuse				
Altair				
Capella				
Denebola				
Procyon				
Regulus				
Sirius				
Spica				
Vega				

Notes

C4

Double Stars and Star Clusters

Purpose

In this exercise, we will observe some of the groupings of stars that we find in our galactic neighborhood. We will observe their color (if any), brightness, and apparent angular size (if any).

Procedure

1. Start *RedShift College Edition*, bring the *RedShift 3* screen to the front, and open your personal settings file. Change the sky settings as shown in the table below.

Date	Today's date
Zoom	0.5

In the Display menu, turn off all items *except* **Deep Sky**, **Stars**, **Constellations**, and **Guides.**

2. In the data table on page 346 is a list of double stars and star clusters. Choose one double star and one open cluster visible at this time of year. From the Information menu, choose **Find** to open the Find window. For the double stars, set the **Find:** box to **Stars, Bayer-Flamsteed**. Find the star name in the list, and double-click on it. Then in the List Links box, click the **Information on...** link, and then click the **OK** button to open the star's information box. For the open clusters, set the **Find:** box to **Deep Sky** and **NGC/IC numbers** for the first cluster and to **Deep Sky, Messier numbers** for the rest. As with the stars, find the object in the list, and double-click on its name. Then in the List Links box, click the **Information on...** link, and click **OK** to open the cluster's information box. For each object, in the information box that appears, click on the **Report** button.

When the Report window opens, look at the horizontal graph. Note the area in green. This represents the times during which the object is below the horizon (not a good time to observe it). The area in white is daylight (also not a good time to observe the object). The area in gray is twilight (you *could* observe then, but it wouldn't be very good). The only good times to observe the object are shown in black. Generally, the very best time to observe the object is when it is highest in the sky. This point—the transit point—is represented by the red line on the graph.

If you're happy with the observing conditions presented by this star or cluster, click the **Print** button to make a printout of this visibility report. If this object is too near or below the horizon at a reasonable time, go back to the main *RedShift* screen, and try the next star or cluster.

When you have selected a likely candidate and printed the visibility report, look at the time indicated by the transit line. Pick a convenient date and time as close to the transit time as possible to make your observation.

C5

3. After evaluating each object, click on the **Close** button to close the Report window. If you have decided to observe this object, you need to make two more printouts. First, click on the **Center** button to center the screen on your chosen star or cluster. Then close the information box, and choose **Print** from the File menu to print this screen. This printout will help you locate the area of the sky where you will find the object.

4. Now change the Zoom to **10** to show the sky as it will look through binoculars. Your chosen object will still be centered on this screen, so you should be able to identify it fairly easily. Again, choose **Print** from the File menu to print this screen. When you have finished this second printout, change the Zoom back to **1**.

Repeat steps 2, 3, and 4 until you have generated visibility reports and sky charts for one double star and one open cluster. When all your reports and charts have been printed, plan star-hop paths to each of your objects. Refer to the section in Appendix D on star-hopping to finish preparing for your observing session.

5. The first clear evening after you have prepared your charts, get out your binoculars, a soft pencil and eraser, and a flashlight covered with red cellophane, and head for a good, dark observing site. Spend a few minutes letting your eyes adapt to the dark and locating any constellations you can. After about 10 or 15 minutes, you are ready to begin observing. Find the jumping-off constellation for your first star-hop and, using the red flashlight to look at your star charts, compare the scale of your printout with the real sky. This will give you some sense about how far away each star or cluster will be. You will also probably have to turn your printout to match up with the real sky.

6. Once you have located the right area for your star or cluster, focus the binoculars carefully (see the section "Observing Tips" in Appendix D),

and begin scanning for your double star or cluster. The double stars should look like twin stars. The open clusters should also be obvious. Some of them are fairly small and concentrated, however, so at first they may appear more like fuzzy spots if your binoculars are not sharply focused.

After you have located your object, study it. For the clusters, try using the averted vision techniques described in "Observing Tips" in Appendix D. Below the data table on page 346 are sketching circles for each of the objects. Treat the outline of each circle as the edge of your binocular view, and sketch the double star or star cluster to scale as it appears in the binoculars. If the cluster, for example, takes up roughly 1/10 the width of the binocular field, sketch it to take up 1/10 the circle's diameter. Finally, note the following details in the Remarks section of the data table:

1) Double stars   Note the separation distance between the stars as compared to the width of the binocular field (1/10, 1/20, etc.). Note any star colors in the Color column. Also note any brightness differences between the stars.

2) Open clusters   Note the approximate diameter of the cluster as compared to the width of the binocular field (1/5, 1/10, 1/20, etc.). Note any distinct star colors in the Color column. Make a rough estimate of how many stars you can see in the cluster.

7. Once you have completed your observations, look carefully at your binoculars. Almost all binoculars have at least two different sets of numbers printed on them. The first set will be something like 7 × 50 or 10 × 50. The first number is the magnification, and the second is the diameter of the main lenses in millimeters. The second set of numbers should include the field of view in degrees (and possibly in yards). Write the field of view in degrees at the bottom of each of the drawing circles on the next page.

Now, multiply your estimates of the double star separations and the cluster diameters (as recorded in the data table) by the field of view you just recorded to get the actual double star separations and cluster diameters in degrees. Write these separations and diameters (in degrees) in the spaces provided next to the drawing circles on the next page.

When you have finished your two observations, attach your visibility reports and star-hop charts to the back of this exercise, and turn it in.

C5

Data Table

Double Stars and Star Clusters			
Object	Colors	Date/Time	Remarks, Separations, & Diameters
Double Stars	*Fall/Winter*		
30 & 31 Cygni			
Sigma 1 & 2 Tauri			
	Spring/Summer		
Epsilon Lyrae			
Zeta Ursae Majoris (Mizar)			
Open Clusters	*Fall/Winter*		
NGC 869			
M45			
	Spring/Summer		
M7			
M45			

Sketching Circles

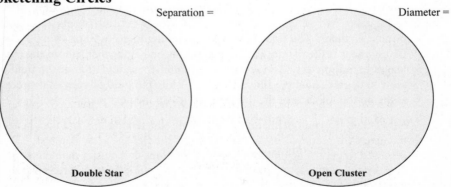

Separation =   Diameter =

Double Star   Open Cluster

Galaxies

Purpose

In this exercise, we wish to actually observe a galaxy. Because other galaxies for the most part are much dimmer and smaller than any of the objects we have yet observed, the observations in this exercise will require a little more patience and understanding. If you are observing in the fall, winter, or early spring you will observe two distant galaxies. If you are observing during the late spring or summer, you will observe our own Milky Way.

Procedure

1. Start *RedShift College Edition*, bring the *RedShift 3* screen to the front, and open your personal settings file. Change the sky settings as shown in the table below.

Date	Today's date
Zoom	3.5

In the Display menu, turn off all items *except* **Deep Sky**, **Stars**, **Constellations**, and **Guides**.

If you are doing this exercise in the late spring or summer, go to Part 6; otherwise, continue through Part 5.

2. In the data table on page 350 are two galaxies that you are to observe. From the Information menu, choose **Find** to open the Find window. Click on the small arrow to the right of the **Find:** box, select **Deep Sky**, and click on **Messier numbers**. Scroll down the Messier list, find each galaxy, double-click on its name, and click **OK**. The sky will center on the galaxy. Click on the galaxy to open its name tag, and then click on the name tag to open its information box. In the information box, click on the **Report** button.

When the Report window opens, look at the horizontal graph. Note the area in green. This represents the times during which the object is below the horizon (not a good time to observe it). The area in white is daylight (also not a good time to observe the object). The area in gray is twilight (you *could* observe then, but it wouldn't be very good). The only good times to observe the planet are shown in black. Generally, the very best time to observe the object is when it is highest in the sky. This point—the transit point—is represented by the red line on the graph.

347

Click the **Print** button to print this report, and then click **Close** to close the Report window. Now click **Close** to close the information box as well. Look at the printout of the visibility report, and find the transit time for the date you wish to observe. Pick a time as close to the transit time as possible to do your observation.

3. When you have evaluated this galaxy, change the Zoom to **1** to get a larger view of the sky. Then choose **Print** from the File menu to make a printout of this screen. This printout will help you locate the area of the sky in which to find the galaxy.

Now repeat steps 2 and 3 for **M33**. When all your reports and charts have been printed, plan star-hop paths to each galaxy. (Refer to the sections in Appendix D on star-hopping and visibility reports to finish preparing for your observing session.)

4. The first clear evening after you have prepared your charts, get out your binoculars, a soft pencil and eraser, and a flashlight covered with red cellophane, and head for a dark observing site. Spend a few minutes letting your eyes adapt to the dark, and locate any constellations you can. After 10 or 15 minutes, you are ready to begin observing. Find the jumping-off constellation for your star-hop and, using the red flashlight to look at your star charts, rotate your printout to match up with the sky and compare the scale of your printout with the real sky. This comparison will give you some sense about how far away each galaxy will be.

5. Once you have located the right area for your galaxy, focus the binoculars carefully (see the section "Observing Tips" in Appendix D), and begin scanning the area, looking for a dim, fuzzy patch that is circular or oval.

    After you have located your galaxy, study it using the averted vision techniques described in "Observing Tips" in Appendix D. Below the data table on page 350 are sketching circles for each of the objects. Treat the outline of each circle as the edge of your binocular view, and sketch the galaxy to scale as it appears in the binoculars. If the galaxy takes up roughly 1/10 the width of the binocular field, for example, sketch it to take up 1/10 the circle's diameter as well. Finally, in the Remarks section of the data table, note any faint details, such as brighter or darker areas.

When you have finished your two observations, attach your visibility reports and star-hop charts to the back of this exercise and turn it in.

6. *If you are doing this exercise in the late spring or summer*, M31 and M33 are not up during the early evening hours. Instead, observe the Milky Way, our own home galaxy. To do so, you will need to find an observing site far away from any lights.

   Begin by preparing a sky chart on which to mark your observations of our own galaxy. From the Information menu, choose **Find** to open the Find window. Click on the small arrow to the right of the **Find:** box, select **Deep Sky**, and click on **NGC/IC**. Scroll through the list, double-click on **NGC 6755**, and click **OK**. You will see the sky move around to center on NGC 6755, an open cluster near the center of the northern part of the Milky Way. It will look like a tiny sprinkling of very faint stars in the center of the screen. Click on the cluster to open its name tag, and click on the name tag to open its information box. In the information box, click on the **Report** button to open the report window.

   When the Report window opens, look at the horizontal graph. Note that the area in green. This represents the times during which the object is below the horizon (not a good time to observe it). The area in white is daylight (also not a good time to observe the object). The area in gray is twilight (you *could* observe then, but it wouldn't be very good). The only good times to observe the object are shown in black. Generally, the very best time to observe the object is when it is highest in the sky. This point—the transit point—is represented by the red line on the graph.

   Click the **Print** button to print this report. Click **Close** to close the Report window, and click **Close** to close the information box as well. Change the Zoom to **0. 5** to shrink the view. Now, choose **Print** from the File menu to make a printout of this screen.

   Look at the printout of the visibility report and find the transit time for the date you wish to observe. This is the time when this cluster is highest in the sky and most easily observed. Pick a time as close to the transit time as possible to do your observation.

7. The first clear evening after you have prepared your chart, get out your binoculars, a soft pencil and eraser, and a flashlight covered with red cellophane, and head for a good, *dark* observing site. Spend a few minutes letting your eyes adapt to the dark, and locate any constellations you can. After about 10 or 15 minutes, you are ready to begin observing. Use the red flashlight to look at your star charts, rotate your printout to

C6

match the sky, and compare the scale of your printout with the real sky.

Once your eyes are adapted to the dark and you have oriented your star chart with the real sky, begin sketching in the Milky Way on your chart. Sketch the brighter areas of the Milky Way with heavier pencil strokes and the dimmer areas very lightly. Note any areas that extend away from the main Milky Way and try to duplicate any darker rift areas in the middle of the Milky Way as accurately as you can.

Once you have drawn the Milky Way, take your binoculars and observe the brighter areas. You will be amazed at how many stars there are in some spots. Sketch in any really bright concentrations of stars on your star chart. Also sketch in any small fuzzy spots you notice in the binoculars. What can you say about the distribution of the clusters compared with the location and extent of the Milky Way?

When you have finished your observation, attach your visibility report and sky chart to the back of this exercise, and turn it in.

Data Table

Galaxies			
Object	Altitude	Date/Time	Remarks
M31			
M33			

Sketching Circles

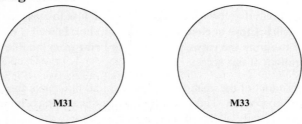

M31   M33

Using *RedShift* as an Observing Companion
Star-Hopping

Introduction and Purpose

The sky is full of distinctive patterns of stars. We call the large ones with which we are most familiar *constellations* and have given them names. But there are lots of smaller, unnamed patterns that are unique and just as remarkable. By using these small, distinctive patterns of stars, you can easily learn to plot a path from the large, easy-to-recognize constellations to just about any place in the sky you might want to go. With *RedShift* along to provide a road map, you can quickly locate identifiable signposts or landmarks to guide you, landmark to landmark, until you arrive at your destination galaxy, nebula, or star cluster. We call this process *star-hopping*, because that's just what we do—hop from star to star and pattern to pattern to find the object we wish to observe. Here's how it's done.

Procedure

1. Start *RedShift College Edition*, and then bring the *RedShift 3* screen to the front. When the *RedShift 3* screen opens, click on the File menu, and open your personal settings file. Then, change the sky settings as shown in the table below.

Date	Today's date
Zoom	3.5

   In the Display menu, turn off all items *except* **Deep Sky, Stars, Constellations** and **Guides.**

   Click on the Filters button, and select **Deep Sky** to open the Deep Sky filters window. Click in the **Labels** box, and choose the type of label that fits the object you are looking for. If it is a number with an M in front, choose **Messier**. If it has the letters NGC in front of it, choose **NGC**. If it has a proper name, such as the Eagle Nebula, choose **Proper names**. Now, adjust the lower magnitude slider (called **Labeled**) to **10**. This slider is used to determine which objects to put labels on. Adjusting this slider to **10th** magnitude will show labels for all objects normally visible with binoculars. Now, click the **OK** button to close the Deep Sky filters window. You will see labels added to help you identify your object, as well as some fainter objects in the display.

D1

2. Once the display parameters are set, locate the object you wish to find. Open the Find window, and select **Deep Sky** in the **Find:** box. If the name of your object begins with the letter M, click on **Messier numbers**; if it begins with NGC, click on **NGC/IC numbers**. Scroll through the list, double-click on the deep sky object you are looking for, and click **OK** to center the object.

   If your object does not show up on the screen at this point, increase the Zoom to **10**, and see if your object shows up. If it does not, continue increasing the Zoom until it does.

3. Then decrease the Zoom to **1** so that you can see a fairly large part of the sky. Find a prominent constellation from which to begin. This constellation should be fairly close to the object you wish to find and should be one you will be able to find in the real sky.

   At this point, move the sky around with the directional arrows on your computer keyboard until the center of the screen is approximately halfway between your deep-sky object and your starting constellation. Then readjust the zoom factor so that your object and the constellation starting point, but not much else, are on the screen. When you have the screen scaled to just show both objects, choose Print from the File menu to make a printout of the screen.

4. Once you have a printout of the deep-sky object and your jumping-off point constellation, work out a star-hop path to your deep-sky object. Look for distinctive groupings of stars, such as straight lines, triangles, double stars, etc., around or between the constellation and the deep-sky object. Sometimes you may even find groups of stars that actually point toward the object you're looking for!

   In general, don't choose star-hops that are very far apart, or they may be difficult to follow in the real sky. If your beginning constellation and the deep-sky object are quite far apart, you may have to zoom in to show enough detail to find easily identifiable star-hops. (If you do increase the zoom factor, however, remember to keep it in the binocular range, around **3.5**.) In this case, you will have to split the distance up into several star charts, hopping from one to the next to the next.

   As you go along, you will find that star-hopping is fun. It is also a very personal thing. As you find groupings that look distinctive to you, you will see some that others will not. We have had students who have seen everything from unicorns to fox terriers to Eddie Van Halen's guitar in the stars! Circle each distinctive grouping of stars (straight line, triangle, double star, etc.), and draw an arrow from each one to the next nearby distinctive group, and then to the next, until you arrive at your deep-sky object.

5. Finally, once you have your star charts marked and ready, you need to determine which night or nights will be best for observing your object and what time of night it will be most visible. This is a job for *RedShift's* Report feature. Instructions for using this feature are found in the "Visibility Reports" section of Appendix D.

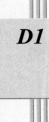

Notes

D1

Visibility Reports

Introduction and Purpose

The Report feature of *RedShift* is a very powerful one for observers. It generates a printout telling you just when the object you wish to observe is most visible. It also lists things like the rising and setting times of your object, times of sunrise and sunset, and the phase and relative angle of the moon.

Procedure

1. Start *RedShift College Edition*, and then bring the *RedShift 3* screen to the front. When the *RedShift 3* screen opens, click on the File menu, and open your personal settings file. Open the Find window, and select the type of object you wish to observe in the **Find:** box. Double-click on the name of the object you wish to observe to select it. Now click **the Information on...** link in the List Links box to open the information box for the object, and click **OK**. In the information box, click on the **Report** button to open the Report window as shown below.

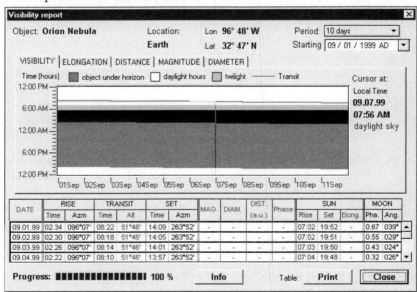

The first report window has a box labeled Period. In this box, you can select the length of time you want the report to cover. If you want to look at the object some time in the next week, for instance, use the default setting of **10 days** to generate a report for the next 10 days. If you want the report to cover a whole month, select **30 days**.

When the computer has finished calculating the visibility of your object over the selected time period, a graph of the object's position in the sky throughout the period will be displayed.

D2

2. Look at the horizontal graph. Note the area in green. This represents the times during which the object is below the horizon (not a good time to observe it). The area in white is daylight (also not a good time to observe an object). The area in gray is twilight (you *could* observe then, but it wouldn't be very good). The only good times to observe the object are shown in black. Generally, the very best time to observe an object is when it is highest in the sky. This point—the transit point—is represented by the red line on the graph.

3. Now look at the area below the graph. This chart lists all the pertinent data available for this object during the selected time period. Here you can find the times of rising, setting, and transit for each day in the period you chose. Only a few days are normally displayed, and you must use the scroll bar to see the rest. Of particular interest is the time of transit. This is the time when the object is highest in the sky and most easily observed. *(All the times shown in the lower section are in 24-hour format. If the hour number is greater than 12, subtract 12 hours from it and make it a p.m. time. If it is 12, make it p.m.; otherwise it is an a.m. time.)* You will want to pick a convenient time as close to the transit time as possible to do your observation.

4. Also notice the column titled **Moon**. This column shows the phase of the Moon and the angle between the Moon and the object you have selected to observe. Since a bright moon often interferes with observing deep sky objects, you'll want to avoid situations where the moon phase is near **1.00** or where the angle between the moon and the object is small. With this printout in hand, you can now plan just what night will be the best for observing your object.

5. You will want to choose a night when the moon phase is the smallest and when the object's time of transit is farthest from sunrise or sunset. Often, the time of transit is during daylight or twilight hours. In these cases, you

will want to get the highest angle of viewing possible in totally dark skies. This can be estimated from the graph.

6. If you are checking on a Solar System object, such as a planet, asteroid, or comet, there are several other options available. The most useful are the **Magnitude** graph (which plots the object's magnitude throughout the chosen period) and the **Diameter** graph (which plots the object's visual diameter over the period). These are accessed by clicking on the **Magnitude** and **Diameter** tabs respectively.

Notes

D2

"Sky Rulers"

As you learn your way around the sky, you will want to be able to estimate angular distances, not only to help others find the things you see, but also to be able to translate star chart information to the real sky. Fortunately, this is an easier task than you might imagine. You have a built-in "Sky Ruler" attached right to the end of your arm—your hand! By simply holding out your hand at arm's length you can make good estimates of angular distances across the sky. Here's how it works.

(1)   Hold up your little finger at arm's length to span an angle of approximately **1°**.

(2)   Hold up your first three fingers together to span an angle of approximately **5°**.

(3)   Hold up your closed fist to span an angle of approximately **10°**.

(4)   Hold up your little finger and your index finger and spread them as far apart as they will go to span approximately **15°**.

(5)   Hold up your little finger and your thumb and spread them as far apart as they will go to span approximately **25°**.

You are probably thinking right now that this can't possibly work consistently because every person has a different size of hand. It is true that these measures are only approximate, but they are a lot closer than you might think. The reason for the unexpected consistency is quite simple.

If your hand is larger than average, your arm is probably longer than most. Or if your hand is small, your arm is likely to be fairly short. So a large hand, held a long way away from your eye, covers about the same amount of sky as a small hand that is considerably closer. Thus, a hand held at arm's length tends to span the same angle for just about everyone.

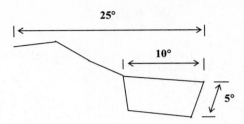

To calibrate your own personal "Sky Ruler," check out the Big Dipper. The distance from the end of the cup to the end of the handle is about **25°**. The distance across the dipper's cup is about **10°**, and the distance between the two "pointer" stars forming the outside end of the cup is about **5°**. With these known angular distances to guide you, you should be able to establish the proper scale for your "Sky Ruler" in no time.

Once you have established the actual distances spanned by your own hand, you can easily estimate things like the altitude of a star above the horizon, the angular diameter of a constellation or the distance from one star to another. Then instead of saying something like, "Look at that star over there. No, not there—over *there!*", you can be a lot more precise and say, "Look at that star about 15° to the north of Arcturus." or "There goes a satellite just 5° below the Big Dipper." It may not sound like a great accomplishment, but knowing how to use your "Sky Ruler" makes observing a lot easier and a lot more fun.

Observing Tips

Binoculars

A pair of clean, properly focused binoculars is indispensable for getting the most out of your viewing sessions. If any of the binocular lens surfaces is dirty, clean it with great care. Get some isopropyl alcohol and some white, soft tissue (Do *not* use facial tissues with lotions or additives.) Apply a small amount of alcohol to the tissue and *very gently* stroke the lens surface radially, from the center to the edge, or vice versa. Do *not* rub hard. Also, try *not* to rub in a circular motion except to touch up problem areas.

To focus the binoculars, pick out a moderately dim star and close your right eye. Move the center focus wheel or bar until the star appears as a pinpoint to your left eye. Now close your left eye, and open your right one. If the star still appears as a pinpoint to your right eye, you are ready to observe. If not, gently twist the barrel of the right eyepiece. Most binoculars are made with adjustable right-side eyepieces. If the right-side eyepiece does not turn easily, do *not* force it, but just leave the binoculars focused as they are. If the right-side eyepiece is adjustable, turn it one way and the other until the right eye also sees the star as a pinpoint. Do *not* readjust the center focus, or you will defocus the image for your left eye.

Now open both eyes. You should see a single, sharply focused image. If you see two images, adjust the spacing between the two sides of the binoculars by gently folding the two sides downward to bring the eyepieces closer together or opening them upward to move the eyepieces farther apart. When both eyes are seeing the same sharply focused image, you will get a lot more out of your observing!

Averted Vision

Have you ever been outside just after the sun went down and the sky was still light? As you are looking up at the sky, suddenly you will catch a glimpse of a star winking at you from the twilight. Your automatic response is to turn your head to look directly at the star, and when you do, it disappears! Then, as you turn away, thinking it was just your imagination, the star pops back into sight again.

This frustrating experience is caused by the structure of our eyes. Our retinas contain two different kinds of light receptor cells. One kind is called a *rod* and the other is called a *cone*. The rods are extremely sensitive and can be triggered by just a few photons of light. However, they do not distinguish between different colors, so they effectively provide only "black and white" vision. Our color vision is provided by the cones. The cones are much more complex cells, and a great deal more light is required to trigger them. The location of the two kinds of cells is also different. The cones are predominantly located in the center of the retina, while the rods dominate the peripheral areas. This means that when we look straight at something, we see in color, but only if the light is fairly bright. Even a slight change in circumstances in our peripheral vision area, however (such as small objects darting at us from the side), would register on the ultra-sensitive, "black and white" rods around the edge of our eye and would allow us to react quickly to these potential dangers.

This helps explain the strange, on-and-off appearance of the first stars you see each evening in the twilight. As you look to one side of the star, its light falls primarily on the rods around the edges of your retina and you see it, even though it is faint. When you turn to look directly at the star, which is dim relative to the rest of the sky, most of the light falls on the cones, and the star disappears. This troublesome phenomenon can, however, be circumvented with a little practice. *To see a dim object, all you have to do is look slightly to one side of it and concentrate on the edge of your vision where the object is.*

As the object "pops" into view, your automatic response will be to turn your eye to look directly at it, but you must resist this urge and keep looking slightly to the side. In this way, you will keep the light focused on the rods and will be able to "see" everything you possibly can. This technique is called ***averted vision***, and it is absolutely indispensable to anyone wishing to do astronomy.

You will find that by using averted vision, you will be able to see all sorts of things that you could not see otherwise. You will be able to detect very faint objects. You will be able to detect faint variations in the objects you look at, such as streaks and markings on planet surfaces or the moon, or dark dust lanes and mottling in distant nebulae and galaxies. If you look directly at these objects, however, all those details will disappear, leaving you with a lot less than you could have seen if you had used averted vision.

Details, Details, Details!

To get the absolute most out of your observing experience, you will need to be very aware of tiny, slight variations in the light coming from the object you are observing. We often look at the beautiful, bright, full-color photographs in our textbooks and expect that all that detail will be just a glance away through a telescope or our binoculars. But, because of the limitations of our eyes, that just isn't so. The details and information *are* there in that faint fuzzy image we see through our instruments, but a good bit of skill and practice are necessary to force our brains to look for those tiny variations and to bring them out.

Move It!

Besides concentrating on your averted vision, there are three other techniques that will help get the last drop of information out of the image you see. The first is motion. Move your eye slightly back and forth, or twitch the binoculars. Motion often triggers your brain to recognize slight differences, and a previously unseen streak or wisp will pop out for just a moment as the image moves.

Memorize It!

The second technique is memory. When you have no idea what you are going to see, chances are you won't see much. Before you go out to observe, take a few minutes to memorize a good black-and-white photograph of the object you will be looking at. Then move away from the photograph and squint at it to simulate the view you will get through the binoculars. Especially note any oddities such as strange symmetries, streaks, mottling, or interesting star formations. If possible, take the photograph along to your observing site. Then as you look through your binoculars or telescope, compare what you see with your memory of the photo. This technique will help you pick out many interesting features you might not have noticed otherwise.

Share It!

The last technique is one of the best and is certainly the most fun. Get a friend or friends to look, and describe to them what you think you are seeing. See if they can see it, too. Often, another person will notice things that you don't and vice versa. By sharing and comparing, we get the most out of what is there.

To Look Is Human; To "See" Is Sublime!

In summary, then, here are the steps to follow in order to "see" everything there is to see. First, use clean, well-adjusted binoculars. Second, look slightly to one side and concentrate on the image. Third, look, look, and then look again. If you think you see some small detail such as a dark streak, a spot, or a brighter area, look away, then look back at the object and see if the feature is still there. Fourth, move the binoculars slightly to bring out any very subtle contrasts. Fifth, memorize interesting details in a photo of the object so that you will know what to look for. And finally, observe with a friend. By sharing with someone else and comparing views, you will both see more, and you will enjoy your experience more, too!

Warp Drive
Efficiency Tips

Use the Control Key shortcuts

Using the <Ctrl> key plus various other keys can help speed up access to frequently used functions such as **Find**. Many other menu items have similar shortcuts that can be found by looking at the item listing when the menu drops down.

Double-click to select

To select an item from a list box, instead of clicking on it once and then clicking on the **OK**, just double-click on the item, and the process is complete! (This does not, however, allow you to change any of the other options in the window.)

Entering data in the Settings Panels

Whenever you have entered any data from the keyboard into any of the settings panels (Time, Location, Zoom, etc.), press the <Enter> key on the keyboard to record the changes. Also note that you can use the Tab key to move from one section of a panel to the next (from the month to the day to the year in the Time panel, for instance).

Open up your windows

When using the *Extras* disk, it is much easier to click on the full-screen button just to the left of the close button at the top right corner of the window than to continually scroll up and down and right and left. Clicking on this button will expand the RSCE window to full-screen size. When finished, just click on the same button to shrink the window back to its original size.

Errors

Currently, there is only one known outstanding error issue. Macintosh users cannot currently play the QuickTime movies contained on the 2nd CD. This issue is under investigation and will hopefully be resolved by the next software release.

Notes

D5

Installing *RedShift College Edition*

Windows:

1. Insert the *RedShift* CD-ROM in the CD-ROM drive of the computer and close the CD-ROM drive door. After a few seconds, the CD will spin, and a small dialog box will open, stating that this application has not been installed on your computer. Click **INSTALL** to begin the installation.

2. The next step asks where you want the software installed. You may install the software on any hard drive in your system, but the default location is probably the best. Click **NEXT** to create a new directory on your C: drive or **BROWSE** to select another location or drive.

3. The next step will install ancillary software needed to run the *RedShift* software. Microsoft's Internet Explorer (version 4 or later) and Apple's QuickTime (version 3 or later) software are necessary to use *RedShift College Edition*. If your computer is new, you might have more recent versions of Internet Explorer and or QuickTime already installed, so watch for such warnings. If so, skip the installation of the software you already have. When you have finished selecting (or deselecting) the appropriate software, click on the **NEXT** button.

4. Now, the *RedShift College Edition* files will be installed. When all the necessary *RedShift* files have been copied to your computer's hard drive, another small window will appear on your screen stating that *RedShift 3* has been successfully installed. Click **OK**.

5. Finally, you will be asked if you wish to install QuickTime 3.0. If you have a higher version of QuickTime already installed on your computer, click **NO**; otherwise, click **YES**. Respond to the prompts that follow as is appropriate for your computer system. When the license agreement is displayed, read the terms of the license, and then click **AGREE** to signify your compliance. Now, another small window will open to lead you through the QuickTime installation.

6. Before finishing the installation, select the **PLAY MOVIE** option to confirm that the QuickTime software is working properly on your computer. This will open a small Movie Player window titled **Sample**. Click on the small VCR-style play button at the bottom of the small window to play the QuickTime movie sample. If the movie plays

E

accompanied by some sound effects, then your installation has been successful. Now close the Movie Player window. When all the QuickTime files have been installed, close the QuickTime file folder.

Macintosh:

1. Quit all open applications.

2. Load the *RedShift 3* CD-ROM into the CD-ROM drive of the computer and close the drive door. After a few seconds, the drive will spin, the *RedShift 3* icon will appear on the Desktop, and a folder window will open. Double-click on the icon marked **RS3 Installer** to begin installation. When the dialog box opens, click **Continue**.

3. The next step asks where you want the software installed. You may install the software on any hard drive in your system, but the default location is probably the best. Click **Select Folder...** if you want to select another location or drive. Once you have selected a location (or decided to use the default location), click **Install** to begin the installation.

4. Now the *RedShift* software files will be installed. When all the necessary *RedShift* files have been copied to your computer's hard drive, another dialog box will open. Click **Quit**.

5. Your Macintosh computer will already have a version of QuickTime installed on it. The *RedShift* CD-ROM has version 3.0 of this software on it. If your QuickTime software is older, you may wish to update it. If you choose to update your files, *you will have to restart your computer when you finish installing these files.* If you wish to update your QuickTime software to version 3.0, continue with step 6, otherwise, you are done.

6. To install version 3.0 of Apple's QuickTime software, look through the folder and click on the icon marked **Install QuickTime...** A dialog box will open to ask if you wish to delete the old QuickTime files before installing the new ones. If you do not wish to delete the old QuickTime files, click **Cancel** to leave your QuickTime installation as it is, and you are done. Otherwise, click **Continue** to install the version 3.0 files.

A Welcome window will open, displaying some information about QuickTime. After you have finished reading the information, click

Continue. A window will open displaying Apple's Software License. After reading the License, click **Agree**.

The next step asks on which drive you want the software installed. Select your hard drive, and click **Install** to begin the installation. When all necessary files have been copied to your computer's hard drive, a dialog box will appear. Finally, click **Restart** to reset your computer so it will recognize the new QuickTime files.

Notes

E

Great Internet Web Sites

Astronomy is perhaps the fastest changing of all the sciences. Almost every day, a discovery is announced that has the potential to change some part of every textbook. Because of this day-to-day nature of our science, learning to keep up with the latest discoveries and developments is essential if you wish to remain knowledgeable for any substantial length of time.

Fortunately, the Internet has made keeping up not only possible, but pleasurable as well. Following you will find a short list of Internet sites that will be valuable resources for you for many years to come. By tapping into these resources, you can get the latest in information, and also download gorgeous, full-color photographs, marvelous movie sequences, and breathtaking 3-D panoramas. With a little practice, the web can become your ongoing window to the universe.

(Please note that due to the nature of the Internet, some of the sites listed may change address or even be discontinued. If you encounter such a problem, please forgive the inconvenience.)

Current Research and Space Missions

The Hubble Space Telescope *http://www.stsci.edu/*

This page is your entry into the incredible research being conducted by the Hubble Space Telescope. In it you will find links to current and past research, ongoing education programs and, of course, all the space telescope photos that have been publicly released, organized by year. If you just wish to see the latest Hubble research, use *http://oposite.stsci.edu/pubinfo/latest.html*

NASA *http://www.nasa.gov/*

This page is your entry into the world of space exploration. In it, you will find links to current and past NASA programs, ongoing public education programs, schedules of future NASA missions and, of course, an archive of photos and movies of NASA missions in the past.

International Space Station Visibility
http://spaceflight.nasa.gov/realdata/sightings/

This page allows you to find the next opportunity for viewing the International Space Station as it goes over your area. Click on the icon labeled "Quick and Easy Sightings by City." Once the sightings page has loaded, find the name of a city near yours, and click on it to find what times you will be able to see the space station in the coming week.

Mars Pathfinder   *http://mars.sdsc.edu/default.html*

On this page, you can keep abreast of the results of the Mars Pathfinder and Surveyor missions. Here you will find links to the results of the Pathfinder and Surveyor research and, of course, an archive of photos, 3-D panoramas, and movies of the surface of Mars.

Galileo Spacecraft   *http://www.jpl.nasa.gov/galileo/*

On this page, you can keep abreast of the results of the NASA Galileo mission to Jupiter and its moons. As in most sites, you will find links to the research and discoveries of the mission, and also an archive of photos and movies of Jupiter and its four large moons.

SOHO   *http://sohowww.nascom.nasa.gov/*
On this page, you can keep up with our changing Sun. Up-to-the-minute images of the Sun are available here along with terrific time-lapse movies of various types of solar activity.

Basic Astronomy

Moon Phases
http://www.ameritech.net/users/paulcarlisle/MoonCalendar.html

This page is a wonderful resource for keeping track of one of the most obvious and easily observed celestial phenomena—the changing phases of the Moon. You can call up any month and year (within reason) and see what the phase of the Moon was (or will be) on each day of that month. By clicking on any day of the monthly calendar, you can also find the moonrise and set times as well as the sunrise and set times (though you will need to reset the latitude and longitude parameters to match your viewing location). *(This is an invaluable aid in planning star parties, field trips or campouts.)*

Views of the Solar System *http://www.planetscapes.com/solar/eng/homepage.htm*
This page is one of the best all-around references on the Solar System. It has sections on the Sun, each planet, comets, asteroids, and the history of space exploration. Each section is a hyperlinked tutorial complete with photographs,

movies, animations, and a hot-linked glossary of terms. Here you will also find the best resources for teaching astronomy. Many of the sections have experiments and demonstrations that can easily be done in the classroom—*an invaluable resource for elementary school teachers!*

Solar System LIVE! *http://www.fourmilab.to/solar/*

This page presents an interactive model of the Solar System. You can see just where each planet is at the time you access the page. From this page, you can also download a freeware program to do similar simulations on your home computer.

Earth Viewer *http://www.fourmilab.to/earthview/*

On this page, you will see a planar representation of the entire surface of the Earth as it looks in real time. You can choose to display recent cloud cover and rain cover, as well as day and night. It's a terrific way to keep up with what's happening on our planet Earth.

The Nine Planets *http://seds.lpl.arizona.edu/nineplanets/nineplanets/*

This page is another excellent reference for information about the Solar System. It has extensive sections on each planet of the Solar System, including multimedia material, and also includes a hyperlinked glossary of terms.

Comets and Meteors *http://comets.amsmeteors.org/*

On this page, you can keep abreast of currently visible comets and meteor showers. Here you will find the most timely information and photos of comets that are currently being observed, links to basic comet definitions, important comet history, and an archive of photos and movies of recent major comets.

Deep Sky Photographs *http://www.ast.cam.ac.uk/AAO/images.html*

This page contains an index to David Malin's breathtaking photographs of galaxies, nebulae, and star clusters taken through the telescopes of the Anglo-Australian Observatory. These photos are generally recognized as the finest available.

The Messier List *http://www.seds.org/messier/*

This page is an excellent reference on the 110 Messier objects. It has a hyperlinked page for each object, complete with an archive of photographs and links to other resources.

F

Daystar—Astronomy, Cosmology and Religion *http://www.daystarcom.org/*

Created by Fred Heeren, author of the book, "Show Me God," this web site explores the ultimate connections between astronomy and religion. Interviews and articles analyze and discuss the philosophical ramifications of current astronomical discoveries. Here you will find interviews with leading astronomers (including Stephen Hawking, Alan Guth, Robert Jastrow, and many others) and theologians regarding the interrelationships between astronomy and religion, as well as a discussion forum open to anyone for participation.

Magazines and News

Astronomy Magazine *http://www.astronomy.com/home.asp*

This page is the gateway to *Astronomy* magazine's web site. Here you will find current astronomy news as well as indexes to information already published or about to be published in *Astronomy*. *Astronomy* is probably the best all-around magazine for anyone with a casual to serious interest in astronomy. Get weekly news stories at *http://www.astronomy.com/story/story_list.asp?intArticleType=2*

Sky and Telescope Magazine *http://skyandtelescope.com/*

This page is the gateway to *Sky and Telescope* magazine's web site. Here you will find current astronomy news, as well as indexes to information already published or about to be published in *Sky and Telescope*. *Sky and Telescope* is a good all-around astronomy magazine and is especially valuable to anyone with a serious interest in astronomy or telescopes. Get weekly news stories at *http://skyandtelescope.com/news/*

Astronomy Now Magazine *http://www.astronomynow.com/*

This magazine features a terrific news service that you can subscribe to without charge. Every day or two, the service will send you an e-mail listing the new discoveries that have occurred during that time period. You can read a very brief summary of the articles or click on the links in the e-mail to go to the *Astronomy Now* web site and read the whole article, complete with photos and illustrations.